FOR PEGGY

Who alone raised Nick for his first four years; kept our house rented and repaired; helped relieve the shortage of nurses in hospitals; coped with gasoline and food rationing; kept me in touch with reality through weekly letters; and, above all, made it so great to come home.

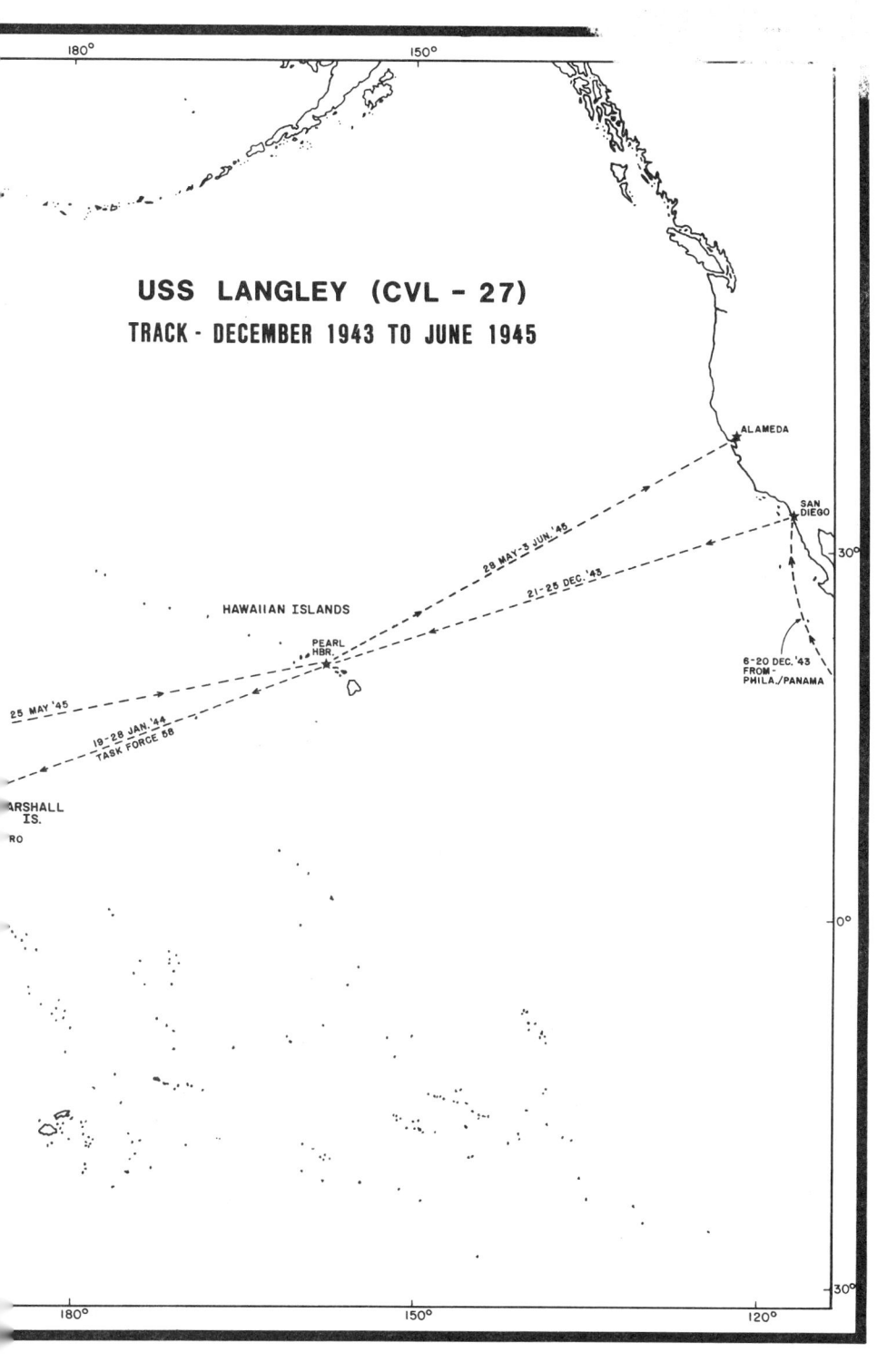

Legend for
USS *Langley* (CVL-27) Operations
from 29 January 1944 to 17 May 1945

①   29 Jan-7 Mar 1944:   Air attacks and bombing of numerous islands in Marshalls; Eniwetok and Majuro used as anchorages.

②   9-13 Mar 1944:   En route Espirito Santo.

③   23-28 Mar 1944:   En route Palau area.

④   28 Mar-1 Apr 1944:   Under Jap air attacks; air attacks on Palau and other Caroline Islands.

⑤   2-6 Apr 1944:   En route Majuro.

⑥   13-20 Apr 1944:   En route area north of New Guinea.

⑦   21-24 Apr 1944:   Air attacks on Hollandia, supported troop landings.

⑧   29-30 Apr 1944:   Air attacks on Truk, pilots down 24 enemy aircraft.

⑨   1-4 May 1944:   En route Majuro.

⑩   6-10 Jun 1944:   En route area west of Saipan.

⑪   10 Jun-6 Jul 1944:   Air attacks Saipan and Rota; air attacks Iwo Jima; under Jap carrier aircraft attack; en route Eniwetok.

⑫   14 Jul-13 Aug 1944:   En route Guam area and return; air strikes Guam.

⑬   29 Aug-9 Sep 1944:   En route Philippine Islands area; air strikes Palau.

⑭   9-25 Sep 1944:   Air strikes Mindanao, Cebu, Panay, and Manila area; return Palau and Ulithi anchorages.

⑮   6-30 Oct 1944:   Air strikes Nansai Shoto, Formosa; Jap air attacks; air attacks Manila; Jap air attacks; air strikes Jap Fleet, return Ulithi.

⑯	1 Nov-24 Dec 1944:	Air strikes Luzon, under suicide attack; attack enemy force; air strikes Manila and return Ulithi; air strikes Luzon, Jap air attacks, return Ulithi; air strikes Luzon, cover landings Mindoro, return Ulithi.
⑰	30 Dec 1944-25 Jan 1945:	Air strikes Formosa, Luzon; air strikes French Indochina, Hong Kong, harbors and convoys; air strikes Formosa; Jap air attack, 100-pound bomb hit forward; air strikes Nansai Shoto, return Ulithi.
⑱	10-28 Feb 1945:	Air strikes Tokyo; air strikes Bonin Island; Jap air attack; air strikes Tokyo, return Ulithi.
⑲	14 Mar-14 May 1945:	Air strikes Kyushu, under Jap air attack; air strikes Okinawa, Jap air attacks; Okinawa landings, cover landings, air strikes; air attacks Jap surface force; air cover Okinawa, support troops Okinawa, Jap air attacks; kamikaze attacks; support troops, frequent air attacks, return Ulithi.
⑳	17 May 1945:	Ship departs for San Francisco.

Lieutenant (j.g.) John Monsarrat, June 1942.

ANGEL ON THE YARDARM:

The Beginnings of Fleet Radar Defense and the Kamikaze Threat

John Monsarrat

Naval War College Press
Newport, Rhode Island
1985

Naval War College
Historical Monograph Series
No. 6

The Historical Monograph Series are book-length studies of the history of naval warfare based, wholly or in part, on source materials in the college's Naval Historical Collection. Financial support for research projects and for printing is provided by the Naval War College Foundation.

Library of Congress Cataloging in Publication Data

Monsarrat, John, 1912.
 Angel on the yardarm.

 Includes bibliographical references and index.
 1. World War, 1939-1945—Campaigns—Pacific Ocean. 2. World War, 1939-1945—Naval operations, American. 3. Monsarrat, John, 1912- . 4. World War, 1939-1945—Personal narratives, American. 5. Langley (Aircraft carrier: CVL-27) 6. Seamen—United States—Biography. 7. United States. Navy—Biography. I. Title. II. Series: Historical monograph series (Naval War College (U.S.)); no. 6.
 D767.M57   1985   940.54'26   85-13850

Table of Contents

Chapter		Page
I	Introduction	3
II	Reporting Aboard at Harvard	5
III	Desk-Bound in the Bureau of Aeronautics	9
IV	Serving on Admiral Towers' and Admiral Nimitz' Staffs	27
V	Learning Radar and Fighter Direction	33
VI	Carrier Duty at Last!	42
VII	Shaking Down the *Langley*	59
VIII	Return to Pearl Harbor	67
IX	The Capture of Kwajalein, Majuro and Eniwetok	70
X	Espiritu Santo and the First Strikes on Palau	74
XI	The Assault on Hollandia	80
XII	Operation Forager and the Battle of the Philippine Sea	85
XIII	Under Halsey to the Palaus and Philippines	93
XIV	The Battle for Leyte Gulf	100
XV	Supporting the Troops on Leyte	111
XVI	The Task Force Meets a Greater Power	117
XVII	Lingayen Gulf and the South China Sea	124
XVIII	The *Langley* Takes Her Lumps	130
XIX	Busy Interlude in Ulithi	140
XX	Raids on Tokyo in Support of Iwo Jima	144
XXI	Fast Carriers Versus Kamikazes at Okinawa	148
XXII	Homeward Bound	174
Chronology of John Monsarrat's Service		179
Appendix		181
Index		185

List of Illustrations

	Page
USS *Langley* (CVL-27) Track, December 1943 to June 1945	i
Lieutenant (j.g.) John Monsarrat, June 1942	iv
Rear Admiral John H. Towers, Chief of the Bureau of Aeronautics, 1942	17
USS *Langley* at cruising speed	58
Downed pilots being rescued by the Vought Kingfisher float plane	83
The survivors are transferred to the submarine *Tang*	83
Officers and enlisted men of the *Langley*'s radar plot seated in front of the ship's scoreboard	116
The *Langley* rolls heavily in the typhoon of December 1944, while a battleship rides steadily astern	120
Radar plot of the suicide raid hitting the *Langley* and *Ticonderoga* off Formosa, 21 January 1945	132
Lieutenant Filo Turner and fire-fighting party on the *Langley*'s flight deck after she was hit by a kamikaze's bomb, 21 January 1945. The *Ticonderoga* burns in the background	135
Air operation plan for the first strikes on Tokyo by Navy planes in support of Marine landings on Iwo Jima	145

I
Introduction

I wish to have no Connection with any Ship that does not sail *fast*, for I intend *to go in harm's way*.
         John Paul Jones, November 1778

This book was written in 1979 as a family memoir for distribution to the author's children and grandchildren. In 1983, at the request of the Naval War College Press, certain additions were made to the text but the substance of the original was not changed.

When the Japanese attacked Pearl Harbor on 7 December 1941, my wife Peggy and I were living in Westport, Connecticut, awaiting the imminent arrival of our firstborn son, Nicholas. I was the vice president of an advertising agency, Platt-Forbes, Inc., of New York City, which was headed by two friends, Rutherford Platt and William A. Forbes. I had had no military training in school or college. However, I had just returned from an extended trip to war-ravaged England where I had visited the squadrons of the Royal Air Force that were operating with engines, propellers, or aircraft manufactured by one of our most important clients, United Aircraft Corporation in East Hartford.

The extensive report I wrote after this trip came to the attention of Rear Admiral John H. Towers, Chief of the Bureau of Aeronautics in Washington. Admiral Towers was one of the first three naval aviators, having learned to fly as part of the Navy's first contract for the purchase of airplanes. A few days before Pearl Harbor, Lauren D. (Deac) Lyman of United Aircraft took me to meet Admiral Towers and to amplify the aspects of my report that interested him.

My account of the USS *Langley* in this narrative takes her from her precommissioning detail to the completion of her wartime cruise. For those interested in her subsequent fate, she was mothballed and placed in storage at the end of World War II; subsequently recommissioned and loaned to the French Navy, in which she was rechristened and served as the *Lafayette*, and ultimately was returned to the US Navy and scrapped.

> John Monsarrat
> Sharon, Connecticut
> July 1983

II
Reporting Aboard at Harvard

On 1 July 1942 Peggy drove me to the railroad station in Bridgeport. Feeling very uncertain and self-conscious in uniform, and laden with whites, blues, and khakis for every occasion, I caught the train for Boston and then commandeered a taxi to take me to the Littauer Building in Cambridge. Throughout that unforgettable day, seven hundred other men joined the long line in front of Littauer to be logged in and assigned to quarters in the historic dormitories facing the quadrangle that was Harvard Yard. Most of the men were in their twenties, with a few in their early thirties. All had attended college. But there the similarities stopped. They came from forty-three states, had attended two hundred thirty colleges and universities, and were drawn from fifty-five professions and occupations. Ten of us were advertising men.

It had taken us all a long time to get there, but once arrived, we were put to work immediately. The Navy's predilection for alphabetizing came into play right away. Strictly according to how our last names were arrayed in the alphabet, we were assigned to platoons, companies, battalions, dormitories, rooms, and roommates. As an "M," I was assigned to a platoon in the sixth company of the second battalion; and as a "MONSA," I was assigned to a room in Matthews Hall and the middle bunk in a three-tiered arrangement, sandwiched in between Adolphe Monosson and Edward Monsour.

We were a disparate but fortunately compatible trio. Eddie Monsour, a soft-spoken Southerner from Shreveport, was the youngest. At twenty-four, he had just completed his master's degree in geology at Louisiana State University and was unmarried. Monosson, at thirty-one was from Brookline and had completed fifteen years in the photo finishing business in Boston. He was a Tufts graduate, bluff, blunt-spoken, but good-natured and possessed of skills which the Navy badly needed for the photographic units on board its aircraft carriers. At twenty-nine, I was the middle man; I had been in the advertising business for seven years, and thanks to having been a small-boat sailor most of my life, I was the only one of the trio who knew port from starboard and forward from aft.

The business of learning naval terminology and idiom was certainly a contributing factor to the shock which many of the men found themselves undergoing. It seemed somehow unreal to be required to refer to a window as a porthole, stairs as a ladder, the ceiling as the overhead, and a wall as a bulkhead.

On the more serious side, both the regimen and the curriculum were very strenuous and demanding. Like most Navy schools all the way from the Naval Academy to the Naval War College, this one was operated on the theory that if you challenged men to learn at a much faster rate than they were accustomed to, they would somehow find within themselves the ability to cope and to achieve. At Harvard, we were expected to acquire a more than rudimentary knowledge of Navy administration and organization, Navy customs and traditions, the elements of navigation, fundamentals of seamanship, some signaling and communications, and the basic responsibilities of an officer of the deck. The course was to last only sixty days but there was a mountain of information to be absorbed, much of it couched in unfamiliar terms, and since most of us were several years away from the routine of academic study, it was hard going.

To guide us through this maze, there was a faculty that consisted mostly of Naval Academy graduates, many of them very recent. They varied widely in their abilities to teach; but I think they were all both interested and amused to find out how much of their own recently acquired knowledge they could communicate to a band of nautical ignoramuses!

On the physical side of our training, we were given a stiff dose of calisthenics at six o'clock every morning, a couple of hours of marching and drilling in formation every day, and a fairly Spartan diet. Most of us came from sedentary occupations and were stiff and sore nearly all the time; and all of us had sore arms the whole time we were there from the seemingly endless courses of injections which the Bureau of Medicine in its wisdom decided we should have. At times, merely raising your arm to salute was a problem, and after the second typhoid shot many of us felt we would rather have the disease than the preventative!

In short order, however, we did begin to toughen up physically, begin to learn a surprising amount about the Navy, and even begin to look like a disciplined military unit. Perhaps more important, we began to *feel* like one.

I can still remember the tingling sensation which would go up and down my spine when, with the entire seven hundred of us at attention in formation in Harvard Yard, the reports would be called out in parade-ground voices up through the chain of command: "Sir, the Sixth Company is formed." "Sir, the Second Battalion is formed." And finally, to the Commanding Officer of the school, "Sir, the Regiment is formed."

Somehow you knew and took pride in the fact that whatever unit you were in, it could not be completely formed unless *you* were there! And in that war all of us *wanted* to be there. Pearl Harbor was the most powerful catalyst of public opinion imaginable and, together with what I had seen and experienced in England, made me eager to take part in whatever could be done to end the menace to our country.

Our basic textbooks were the same as some of those in use at Annapolis: Knight's *Modern Seamanship*, Dutton's *Navigation and Nautical Astronomy*, *The Bluejackets' Manual*, *The Watch Officer's Guide*, and Lovette's *Naval Customs and Traditions*. Taken together, they formed a very solid basis on which to build.

All of us had been commissioned provisionally. My own designation was AV(P), standing for Aviation Volunteer Provisional. Upon my completing the course at Harvard satisfactorily, my designator would be changed to AVS or Aviation Volunteer Specialist. What I was to specialize in remained a mystery. The orders of most of my classmates did not go beyond the school at Harvard, and they would not know until graduation where they would be sent next. Unlike theirs, my original orders specified that, upon graduation, I was to report for duty in the Bureau of Aeronautics in Washington, but there was no clue as to what my job there would be. As we approached graduation, one by one my classmates received their new orders; some were to go directly to ships or shore stations, but most to Navy specialist schools such as the Mine Warfare School, the Amphibious Warfare School, the Communications School then being started at Harvard, and the Air Combat Intelligence School at Quonset in Rhode Island. It seemed to me that I would have been a logical choice for the last-mentioned school, but at least in the Bureau of Aeronautics I would be in the aviation branch, and perhaps that could somehow lead to aircraft carrier duty, which is what I wanted most. There remained in my mind, however, a nagging concern that someone would try to get me into Public Relations because of my background in advertising. If I were going to be a naval officer, I wanted to avoid that on the grounds that it would be too much like being a civilian in uniform.

Toward the end of August we took our final examinations in all the various subjects, were given our graduation certificates, and detached from the school to join the Navy in earnest.

I think we all left Harvard with a new respect for how much the Navy could teach us when it had the time to do so, an impression confirmed when we attended more advanced schools later in our careers. With only a brief stopover in Westport to see Peggy and Nick, I left to test the waters in Washington.

III
Desk-bound in the Bureau of Aeronautics

In those pre-Pentagon days, the Navy Department was on Constitution Avenue, where it occupied a very long, low, "temporary" building that had been put up in World War I and was known as Main Navy. Extending for what seemed like miles in each direction, and three or four decks high, the building had three entrances, all from Constitution Avenue. These gave access to eight long corridors along which were arrayed the offices of the various bureaus into which the administration of the Navy was divided. At right angles to these corridors, a transverse corridor ran along the Constitution Avenue side of the building and housed the offices of the secretary of the navy, Frank Knox, and the chief of naval operations, Admiral Ernest J. King. The last corridor at the west end of the building housed the Bureau of Aeronautics, BuAer in Navy parlance, still headed by Rear Admiral John H. Towers.

When I reported to the duty officer, he logged me in and showed me how to find the personnel officer who would brief me on what part of the huge bureau I was to work in.

Here came a real shock. My job was to relieve Joy Hancock so that she could get into uniform and take over a top job in the Waves, which was in the process of getting organized. If I had not known Joy and how capable she was, I'm sure I would have resented going through the long commissioning process merely to relieve a civilian woman employee. However I did know her well, and knew also how much she was respected in the senior echelons of the aviation branch. So, off I went to see her in her office on the main deck, a few doors down the corridor from the admiral's office.

Joy was her usual effervescent self and looked like the cat that had just swallowed the canary.

"I thought I'd be going back to England, or perhaps to Air Combat Intelligence School," I said. "Who put the finger on me?"

"I did," said Joy. "When the admiral approved me for this new job in the Waves, I was to find my own relief, and when I saw your name on the list for Harvard, he agreed to my request that you be ordered to report here directly from school. So, my friend, we've got a lot of work to do!"

All I knew about Joy's job was that she had the title of Director of Editorial Research, and that she was the person who approved or disapproved advertising copy pertaining to aviation products manufactured under Navy contracts. As she began right away to explain it to me, I became aware that the job included a whole lot more.

Organizationally, the Editorial Research section was part of the Office of the Chief of Bureau, and its director was authorized to sign correspondence for the Chief of Bureau "By Direction," a responsibility which even I knew must have awesome penalties for misuse! In addition to reviewing for security all aviation advertising copy submitted to it by the Navy's suppliers, the Navy Department's Public Relations Branch frequently referred newspaper and magazine articles to the office for factual accuracy and security with respect to naval aviation. Both of these assignments required a thorough knowledge of what had already been published in the aviation trade press, a detailed knowledge of naval aircraft, and a knowledge of where to go in the bureau to get factual information beyond one's own experience. Since I had subscribed to the leading American and British aviation trade papers and written advertisements and articles for them, I at least had a start on these aspects of the job.

Another part of the job was to write speeches for the Chief of Bureau as he might require them, and to prepare the annual summary on naval aviation for the *Encyclopaedia Britannica*.

All these assignments were closer to Public Relations than I wanted to come, and I was glad to learn that there were others of a different nature. One was to edit and produce the Bureau of Aeronautics Newsletter, a restricted official publication put out bimonthly and distributed to the naval officers in the bureau and the wardroom libraries of all aircraft carriers and naval stations. Its purpose was to share with interested personnel the developments in the bureau and in the fleet that could be reported under a "restricted" security classification, as opposed to the more stringent "confidential" and "secret" categories.

We officially approved and kept records of all squadron insignia. In the safe in our office, we kept the engineering performance records of every type of airplane in Navy service: its maximum speed, service ceiling, range, payload, rate of climb, etc. We also kept a file on the attrition of naval aircraft and updated it on the basis of dispatches received from ships and naval air stations all over the world. Finally, a whole series of ad hoc assignments was given to Editorial Research, most of which proved to be extremely interesting.

The office consisted of two rooms, a small one for the director and a fairly large one for three experienced civilian employees, all women, and the files. The two were connected and both opened onto the BuAer corridor through a pair of latticed swinging doors, much like those that Hollywood loves to depict at the entrance to old Western saloons.

Joy spent about a week taking me through the details of the assignment and introducing me to the officers in other parts of the bureau with whom I would be doing the most business. Then, when I was ready to relieve her, or rather when she thought I was, she arranged with the admiral's aide for me to join a small group of reporting officers in paying a courtesy call on Admiral Towers.

The aide briefed us and warned us that the admiral was very worried about the lack of government living quarters for new arrivals, who had to scramble as best they could to find a place to stay in the turmoil of wartime Washington. "The Admiral is bound to ask you if you've found a place to stay," said the aide, "and whatever you do, don't tell him you haven't. There are always park benches!"

The admiral was courteous but brief in his welcoming remarks, and we were in and out of his office in a very few minutes. He did ask us about housing, and we all said we were making do. So, now I was officially the officer in charge of editorial research, a strange title for a section that seemed to do a mixture of things, few of which bore any relation to editorial research.

To help resolve my own family's housing problem, I combed the Washington newspapers and got the names of a few rental agents from newly made friends in the bureau. Peggy came down from Westport to help in the search, and, as luck would have it, we learned from one of the agents that a congressman from Wyoming was interested in renting his house in Falls Church for at least a few months. We got its address and drove out immediately. As we pulled up in front of it, a car backed out of the driveway and in it were the congressman, his wife, and children about to take off for Wyoming. They stopped long enough to show us through the house, and we took it on the spot, happy to have a nice, two-bedroom place in a good neighborhood and convenient to the bus lines into Washington. Peggy went back to Westport to rent our house there, and we soon moved Nick and our baggage to Falls Church.

Nearly a year after Pearl Harbor and two years after the Battle of Britain, I was surprised to discover how much jealousy and friction there was within the Navy Department between the "battleship admirals," representing the surface warfare forces, and the "air admirals," representing the aircraft carrier forces. Just as in civilian life the proponents of air transportation saw themselves as crusaders against the hidebound restrictions of the railroads, so had the early proponents of naval air power like Admiral William A. Moffett and Admiral Towers seen themselves as crusaders against the notion that naval air was nothing more than an adjunct to the Battle Force. What was amazing was that they were *still* having to do so! Originally, the allocation of scarce funds between the surface ship, submarine, and aviation branches of the Navy was the root of the problem. But now money was not an issue; there was ample for all. Yet the friction continued, and one of the battlegrounds was in the area of public relations.

Admiral King took a very dim view of all public relations activities. It was said that, if he had his way, he would hold only one press conference and that to announce at the end of the war, "We won." Nevertheless, he recognized that there had to be some sort of public relations organization to handle the press and the massive public concern about the Navy in the wake of the debacle at Pearl Harbor. Consequently, the Navy Department had a Public Relations section which controlled all the official communiques and press releases, and had cognizance of what could and could not be published or broadcast under wartime restrictions. This section was headed by a captain who at first did not have a single naval aviator on his staff.

BuAer and many of the aviators in the fleet felt that Navy Public Relations was giving naval aviation the short end of the stick, either deliberately or through ignorance. By the time I arrived on the scene, BuAer had succeeded in getting Public Relations to refer to the bureau for comment all material that dealt with aviation subjects before releasing it. Through this process we were able to check the accuracy of many magazine articles dealing with naval aviation and help the authors to correct misstatements. My office was on the receiving end of the inquiries which came over from Public Relations, and I was authorized to go anywhere in the bureau for help when the accuracy of a statement was beyond my competence to judge. It then became my responsibility to sign for Admiral Towers "By Direction" the statement, "The Bureau of Aeronautics has no objection to publication."

Another improvement in the system from BuAer's point of view was the willingness of the Public Relations branch to add to its staff a naval aviator in the person of Lieutenant Ben Hoy. Although Ben had lost a leg in an accident, he was permitted to serve on active duty, and his new post at least gave naval aviation a voice at a focal point in the flow of information. Being the only aviator in Public Relations, Ben had no one with whom he could indulge in aviation talk and often walked the whole width of the building to visit BuAer. We became good friends and frequently lunched together to compare notes.

Two other officers in Public Relations who were helpful in keeping an eye out for the interests of naval aviation were Ensigns Frank Rounds and Joseph Pulitzer. Frank had been a reporter for *U.S. News and World Report*, and Joe had begun to work with his family's *St. Louis Post-Dispatch*. Though not experts in aviation affairs, both were very able and very bright, and were in no way hindered by prejudice one way or the other in the squabbles over parochial interests within the Navy.

I soon learned, however, that there were other forces at work behind the scenes. Lieutenant Commander H.B. "Min" Miller had been brought into the bureau from a patrol plane squadron to head up a section known as Training Literature. Its responsibilities included the preparation of all the training manuals and training films for the aviation branch, which it did extremely well, thanks to Min's ability to assemble a staff of competent officers from the ranks of publishing and the graphic arts. Among the members of the group whom I came to know were Fred Tupper of *New York Times* fame; Robert Osborne, the cartoonist who created the marvelous character "Dilbert," a mythical naval aviation cadet who did everything wrong and became a legend in posters at Pensacola and Corpus Christi; and Edward Steichen, already a legend as one of the world's greatest photographers.

Busy as Min was in fulfilling his section's assignment, he nevertheless became the spearhead of a sub-rosa publicity department for naval aviation as a whole, whose objective was to combat what was seen at the time as apathy or worse on the part of Navy Department Public Relations. He and his talented group never lost an opportunity to encourage stories favorable to the cause of naval aviation, to interest civilian writers in pursuing them, and to fight for recognition of the contribution that naval aviators were making to the effort on both oceans. In retrospect, it is hard to understand why such effort was required in view of the facts. Yet, at the time, I became aware that naval aviators, in their distinctive green uniforms, returning from the fleet on leave were sometimes mistaken for hotel doormen. Regardless of the causes within and without the Navy Department, there was a need for Min's group.

I first learned of this aspect of Training Literature's business when it asked me to write for the magazine *Flying* an article summarizing the accomplishments of naval aviation in each of the major actions against the Japanese in which it had played an important role. On investigation, I found that Min had agreed with the editor of *Flying*, Max Karant, whom I knew slightly, that if Max would devote an entire issue of the magazine to naval aviation, he would provide a series of articles by the officers in the Bureau of Aeronautics responsible for the procurement of aircraft: fighters, patrol planes, dive-bombers, torpedo planes, observation scouts, and lighter-than-air. The officer in charge of each of these projects, or "desks" in BuAer parlance, had agreed to prepare an article on his specialty, and I was to write the lead story on how all these component factors came together in combat. Each author was to be identified at the outset of his article and his photograph was to accompany his by-line.

To say that I had mixed emotions about this assignment was to put it very mildly. On the positive side, it would be extremely interesting to research and write the story of naval aviation's participation in the battles to date, notably Guadalcanal, the Coral Sea, and Midway. But on the negative side, I felt that it was improper for the Navy to favor one magazine by giving it such extensive and exclusive assistance with its editorial responsibilities. Furthermore, having dealt with all the aviation magazines of the day in my role at the Platt-Forbes advertising agency, if I were going to grant such privilege to one of them, *Flying* would have been the least likely to receive it. It seemed to me the least authoritative of them all.

At home in Falls Church that night I was undoubtedly very bad company for Peggy because these issues bothered me inordinately and I knew I had to decide right away how to handle them. The "request" from Training Literature was really an order, and I knew I had to comply. But by the time I reported for work at eight o'clock the next morning, I had hit upon a plan to salve my own misgivings: my contribution would be anonymous and no photograph would be attached.

The Navy had a system whereby every unit in combat, whether individual ship, squadron, task group, or task force, was required to submit an action report afterwards. These reports were all on microfilm in the basement of the Navy Department. During the following week, I read the action reports of the major units at Pearl Harbor, and those of the aircraft carriers and individual squadrons involved in the Coral Sea, Midway, and lesser actions that had taken place up to that time. Based on copious notes taken from all the reports, I wrote the article requested and delivered it to the project officer in Min Miller's office. A week or so later, he called me to say that he had submitted it to the censors in Public Relations and they were angry about the account I had given of a submarine sinking. I was to report immediately to a captain in Naval Operations and explain why I had breached the rigid rule against revealing when and how enemy submarines had been sunk. When I presented myself to the captain, I spent an uncomfortable fifteen minutes on the receiving end of a spectacular tirade. He made it clear that the less the enemy knew about which of his submarines we had sunk, and where and how, the more difficult we made it for him; and vice versa. When he finally wound down, I asked him if I might explain why I had included the incident in the article. Reluctantly, he agreed, and I told him that the submarine in question was tied up to the dock at Japanese headquarters in Jaluit, where it was sunk in full view of all the Japanese at the base. Therefore, they already knew where and how and when the submarine had been sunk, and the only people who did *not* know about it were the American public. "Young man, don't give me an argument. An order is an order, and the order is not to give out any details on submarine sinkings. Now take it out of the article and get out of here!" In the Navy, regardless of logic, four stripes always beat one and a half!

The rest of the article came through unscathed. As the various pieces of the *Flying* project came together, I got a call from Training Literature asking where my photograph was and pointing out that I had not signed the article. I explained that I preferred not to sign it and not to have my photograph connected with it. I was afraid this might get me in trouble with Min, but I never heard anything more about it; and the special issue finally appeared.

Not long after I had settled in at Editorial Research, Admiral Towers' aide called me into his office and said the admiral wanted me to write a speech for him to give at the Grumman Aircraft factory in connection with the presentation of a Navy "E" to Grumman for its excellence in performing as the prime contractor on the new Navy fighters and torpedo planes. He gave me not a single clue as to what the admiral wanted to cover in his speech; he said only that I should have it ready three days hence and typed on pocket-size cards in large print.

This was my introduction to the peculiarity of the Navy staff system, according to which junior officers are expected to read the minds of their seniors, and work in the dark toward a solution to a problem. If you do the job satisfactorily, it is accepted. If not, it is bounced back, sometimes without any explanation as to why it has been rejected, to be done over again. A strange way to work! Fortunately, Admiral Towers always accepted what I wrote for him, and as far as I know he delivered the speeches as written.

When it came time to rewrite and update the history of the bureau for the *Encyclopaedia Britannica*, I was able to draw on the copy for previous editions as well as on the research I had done for *Flying*. I enjoyed doing the article and was pleased that it passed through the approval process without any changes.

One of the writing assignments which came to me from the admiral was truly sad. During the Battle of Midway on 4 and 5 June 1942, an entire torpedo squadron from the *Hornet* was shot down while attempting to make a long-range attack on the Japanese carriers. Led by Lieutenant Commander John C. Waldron, all fifteen of the antiquated Douglas dive-bombers from Torpedo Squadron 8 made the long flight, found the enemy with his attack planes rearming and refueling on deck, only to be themselves destroyed by thirty Zero fighters awaiting their arrival. Only one of the American planes got close enough to attack, and its torpedo got hung up in the release mechanism at the crucial moment. Of all the pilots and their radiomen, only one survived: Ensign George Gay, who floated underneath a seat cushion after his plane hit the water while the whole Japanese task force steamed past.

At the time of the battle, the *Hornet* had on board one of the first of the combat photography teams that Min Miller's group was organizing under Edward Steichen's guidance. This one included John Ford from Hollywood, who photographed the takeoff of the entire squadron, plane by plane, as it rolled down the flight deck in the mid-morning sunlight. With their cockpit canopies open for takeoff, the pilots were clearly recognizable, and it was heart-rending to see them go; all the

Rear Admiral John H. Towers, Chief of the Bureau of Aeronautics, 1942.

more so because it was widely believed that the men knew that they were going to fly beyond the range at which they could hope to return to the *Hornet*.

The film was in color and had been beautifully edited to include footage of the carrier herself; to superimpose under each takeoff scene the names of the pilot and radioman; and to conclude with a closeup of the Navy Cross and the citation that was awarded to every participant.

The assignment given to me was to devise the best way to transmit a sixteen-millimeter print of the film to the next of kin of each member of the squadron. After I had seen the film, I felt that it was just too strong and sad a memorial for the recipients to receive without warning. So, without consultation, I wrote a letter for the admiral's signature to each of the next of kin and to Ensign Gay, telling them of the existence of the film, describing what it contained, and offering to send them a print of their own if they would like to have one. The admiral signed all thirty of the letters, and, as we expected, every family promptly replied that they would like us to send the print. I then wrote them a letter saying that their print was on the way; and shortly thereafter the most touching letters of appreciation began to arrive. While it was heartening that the Navy had handled the matter with compassion, it made me sick to think that our country had let its aviators down to the extent of not providing them with modern aircraft capable of meeting the enemy on at least an equal footing. If Congress had listened to men like Gene Wilson of United Aircraft, these heroic men would not have had to fly in slow, obsolete airplanes. Their loss, however, was not entirely in vain: because of the attack, the enemy carriers could not finish refueling their aircraft before the dive bombers arrived, and all were sunk by the end of the battle.

In the course of a routine day, an avalanche of mail would descend on my desk. I am not sure whether speed-reading systems had been invented in those days, but certainly I had never been exposed to them. In self-defense, I taught myself to be such a rapid scanner that, in a kind of administrative triage, I could sift through a pile of mail very quickly and separate it into three batches: one that could be handled routinely by my staff of civilians, one that I could handle without information beyond my own files, and one with which I would need help from other sources in the bureau.

The mail in the last-mentioned batch necessitated my finding out who would have the right answer to a question and then tracking that individual down and getting some time with him. Since the right person often turned out to be a fairly senior captain, I met many of those who later on played major roles in the Pacific. Among them were Captain Arthur W. Radford, Captain Ralph Davison, and Captain Harold B. Sallada, all of whom were to command carrier task groups as rear admirals, and Captain Alfred M. Pride who was a kindly mentor to me and who was soon given command of the light aircraft carrier *Belleau Wood*. Another most interesting officer was Commander Frank W. "Spig" Wead, a naval aviator from the Annapolis class of 1916. Spig was an exceptionally competent flier who had been crippled for life in a fall downstairs in his home, and had retired in 1927. He started a very successful comic strip featuring a naval aviator and then went into filmmaking in Hollywood. At the outset of the war, he was reinstated and restored to active duty in operational intelligence in the Bureau of Aeronautics. Admiral Towers and the other old-time aviators had great respect for Spig's intelligence and abilities. His office was close to mine and many times, long after hours, I would see him making his tortuous way out of the building with the help of two canes strapped to his arms.

Late in October, just as I was beginning to feel a bit more comfortable in the job, we received the news that Towers was being promoted to vice admiral and would be leaving the bureau to take a new job at Pearl Harbor with the title of Commander, Air Pacific, or ComAirPac. I was not aware of the long-standing dislike that Admiral Towers and Admiral King had for one another. The friction between them was intensified by the fact that Towers undertook to make known to higher authority the discontent of the aviators, who felt that too many of the top commands were going to nonaviators and that the carrier forces were being mishandled because they were often subject to admirals who had no experience in aviation. There was much substance to these criticisms, as was painfully apparent at Guadalcanal and elsewhere. Towers took every opportunity to try to get the situation corrected and, because the Bureau of Aeronautics reported by law to the Secretary of the Navy and not to Admiral King as CominCh, he could make waves in high places without having his head chopped off. King saw the opportunity to get Towers out of Washington and took it. He stubbornly resisted, however, every effort Towers made to be given command of a fleet or task force, and kept him ashore at Pearl Harbor until the very last days of the war.

As the time approached for the admiral's departure, I asked his aide to put me on the list to say goodbye in person. When my turn came, I congratulated him on his third star and, for the first time since I had reported for duty, he seemed to unbend just a little. Neither of us had ever made reference to our long talk after I returned from England, and he had been formal and militarily correct each time I had seen him since. Now I thought I saw a twinkle in his eye, and I decided to press my luck. "Admiral, when you get out there, you could do me a great favor if you would." "What would that be?" he asked. "Sir, you could send for me. I've been a sailor all my life and I've spent all my adult life working in aviation. I want to get to sea on a carrier!" For a moment I thought I had gone too far, but then he smiled and said something to the effect that he would keep it in mind. On that we parted, and I was glad that at least I tried!

The new chief of the bureau was Rear Admiral John Sidney McCain, fresh from sea duty in the Pacific. No two men could have been more different in manner and in style. Where Towers was always perfectly groomed, McCain looked as if he had slept in his uniform for three nights running. Where Towers was formal, McCain was the height of informality. Joy Hancock took three new Wave officers into his office to introduce them, and his first words were, "Where you from, kiddo?" After I was introduced, he asked where my office was and whether I had a coffee pot. When I told him I was just a few doors down the corridor and Mrs. Marshall, the senior civilian in Editorial Research, made good coffee all day, he said, "Fine. I'll be down and give it a try." Thereafter, he dropped in unannounced nearly every day to have a cup with us and "bat the breeze" with Mrs. Marshall. One of his best friends was the actor Adolph Menjou, and when Menjou came to Washington for a visit, the admiral simply disappeared. Not even his aide knew where he was until he called in to see what was going on. Yet, with all his informality and his striking resemblance to Popeye the Sailor, he seemed to be very much on top of things, and there was no doubt that Admiral King liked him and had handpicked him for the job.

From time to time I saw Joy in the corridors and twice she dropped into the office to share with me an especially juicy morsel of news. On the first occasion, she was returning from Admiral King's office to which she had gone in fear and trembling after receiving a summons with no explanation of why CominCh wanted to see her. "John," she said, "you won't believe it but the old boy actually kissed me!" What had happened was that someone had told King that Joy was a yeomanette in World War I and was now a Wave lieutenant, and he had decided to pin a World War I victory ribbon on her uniform in person. It was so out of keeping with King's reputation for austerity that Joy simply could not believe it. "And right on the mouth, too!" she said with her blue eyes sparkling.

A few weeks later in she came again and started to giggle before she crossed the threshold. She had been working on the design of a new Wave barracks to be erected somewhere, and had just got Captain Davison's approval signature on the large bundle of blueprints she was carrying. "John, I've *finally* made a lasting niche for myself in naval history!" "What have you done now, Joy?" "I'm going to go down in history as the gal who brought bidets to the United States Navy!" And what a "gal!" After the war she was promoted to captain and became the commanding officer of all the Waves.

With the coming of winter, I received visits from two widely different delegations of foreign officers, one British, the other French.

Well in advance of the arrival of the British, I had been briefed on their mission and on what was expected of me in connection with it. All three members of the delegation were aviators in the Royal Navy. Their superiors in British naval aviation had long suffered from being treated as the stepchildren of the Navy, and they were determined to do something about it. We thought we had troubles with our battleship admirals, but their Fleet Air Arm's difficulties with their Navy and with the Royal Air Force were many times worse. Within their own service the aviators had very little to do with planning, and when it came to the design of new aircraft carriers, they had to fight their way through a wall of ignorance. For a long time, the Royal Air Force procured all the Fleet Air Arm's aircraft, and the RAF had no experience in shipboard operations. Virtually none of the top commands in the Navy were given to aviators; proportionately, even fewer than in the US Navy. Not a single carrier was commanded by an aviator. And British naval aviators looked with envy at what they saw as the autonomy of the Bureau of Aeronautics in the United States. In peacetime, funds for research and development of new aircraft had been virtually nonexistent, and were still far short of adequate. The antiquated Douglas TBD torpedo plane with which our Navy pilots went to war was a paragon compared with the Fairey Swordfish, its Royal Navy counterpart. Antisubmarine patrols and over-ocean scouting from shore stations were being conducted entirely by the Coastal Command of the RAF instead of by the Navy, which had primary responsibility for convoy protection.

These and similar complaints had made their way up through the chain of command to Winston Churchill, who gave approval for this study group to come to the United States and learn how and why the Bureau of Aeronautics had come into being with a view to reexamining the situation in Britain. My job was to help the group in any way possible.

I was thankful that I had been briefed in advance, because I had never studied the origins of the bureau beyond the once-over-lightly I had given them for the *Encyclopaedia*. I now burrowed into the subject much more thoroughly and by the time the Brits appeared, I was much better prepared to point them in useful directions.

We hit it off very well right from the start, partly, I suppose, because of my visit to England and Ireland for United Aircraft. They went into considerable detail on their problems, even to the point of discussing their resentment over the fact that the Royal Navy would not allow them to wear their wings over the left breast pocket of their uniforms, as was the case in their own RAF and every other flying service in the world. They were required to wear a very inconspicuous version of them on their sleeve like a line officer's star. And when it came to discussing the problems they had in getting carriers designed properly, one of them told me an amusing experience he had had.

He was called in to look over a set of the naval architect's preliminary plans for the hangar deck of a new carrier. After studying them for a few moments, he turned to the Royal Navy nonaviator in charge of the project and said, "I don't see any provision here for the stowage of spare propellers. What do you intend to do about that?" There was complete silence for a whole minute. Finally the light dawned, and the RN officer said, "Ohhh . . . you mean *airscrews*!" The subject had never occurred to him!

I described our own Navy's involvement with aviation from its beginning with the Wright brothers, through the controversy with Billy Mitchell and the Army, to the efforts of Admiral Moffett and Admiral Joseph M. Reeves to gain recognition and funding for naval aviation. The establishment of the Bureau of Aeronautics in 1921 was probably the most significant step along the way, and I had ready for the British a transcript of Admiral Moffett's testimony before Congress in support of the formation of the bureau, along with the text of the act of Congress that finally established it. The latter specified that the bureau be under the overall cognizance of the *civilian* head of the Navy in order to protect it as much as possible from antiair bias within the military establishment.

This material was pay dirt for the British officers, and I was able to cite library references through which they could learn more details and pick out the facts that would best serve their cause at home. As we wound up our meetings, which lasted several days, I cautioned them not to think that everything was a bed of roses in our own situation, citing the resentment of our aviators over the fact that aviators were not assigned to major fleet commands, and their conviction that air admirals, if given the chance, would make much better strategic use of the carrier task forces than did surface warfare admirals, colloquially known as the Gun Club.

All in all, this interlude was a very pleasant one for me. I felt that I had learned a lot and also that I had been helpful in giving them a much better understanding of our attacks on problems very similar to theirs. We parted good friends and I have always been curious as to how they fared on their return to England.

My visit from the French delegation was a horse of a totally different color. One day, out of the blue I was summoned across the corridor to Captain Sallada's office where he and another captain were conferring on plans to receive a visit the next day from a group of Free French officers arriving from England. I sensed at once that there was considerable tension in the room. I was brusquely informed that the French were coming, and that I was one of the officers in the bureau on whom they would call. Nothing was said about the purpose of their visit; only that they had just arrived from General de Gaulle's headquarters in London. They would be asking me questions about the performance characteristics of various US Navy airplanes. Under no circumstances was I to answer their specific questions, I was to stick to generalities. Most particularly, I was forbidden to show them any of the performance data I had in my safe. In short, although the captains did not say it in so many words, I was to act polite but dumb. I was rather abruptly excused and went back across the hall a very puzzled j.g.

When the officers arrived at my door the next day, I seated them in a semicircle in front of my desk and equipped them with coffee in the standard Navy manner. After the amenities had been exchanged, they asked me if I was familiar with the purpose of their visit. When I told them I was not, they said they had come to Washington to learn the detailed characteristics of all American naval aircraft in order to determine which types and models might fit in best with their plans for rebuilding the Free French Navy. They had with them a list of the types in which they were most interested, and wanted me to go over it with them, type by type, and provide the specifics for each plane.

I was hard put to it to evade their questions. Their English was too good for me to pretend that I didn't understand them. Six feet behind them, in the safe in the corner, I had exactly what they were looking for. The only way out seemed to be to pretend that I didn't know the answers. From my knowledge of all the aviation trade journals, I did have a good understanding of what had already been published, and was, therefore, in the public domain, available to all. So I went into a long discussion of generalities from the trade papers, knowing full well that they wanted me to be a whole lot more specific. After listening for a while, they began to get angry and made it plain that they wanted a lot more meat and a lot less fat. After an extremely uncomfortable half-hour of further sparring, I told them, "I am not an aeronautical engineer, and I can't answer your questions." At this the meeting broke up, and they went off.

In no time I was summoned back to the captain's office. "I see they've left. What did you tell them?" "Nothing, sir, that they couldn't read for themselves in *Aviation* or *Aero Digest*." "Very well. That's all."

I was really embarrassed and angry to have been put in such a spot. "It's a hell of a way to run a war," I thought, "to give an ally this kind of run-around. And a hell of a way to treat a j.g., to send him into that kind of meeting without knowing what it was all about!"

Not until I read the history of Roosevelt's and Churchill's relationship with de Gaulle, long after the war was over, did I understand the reasons. These officers had come at de Gaulle's insistence. Neither Roosevelt nor Churchill trusted him, and they were not about to give him any secrets, lest they wind up in Nazi hands via Vichy France. Yet, outwardly, because of the help he might be able to give later on, they had to keep up the fiction that de Gaulle was a meaningful ally. They approved the French officers' visit, but at the same time sent word down the chain of command not to cooperate. The captains in the bureau were under strict orders and had put me under strict orders, but were not free to tell me why. A pretty heavy load for a j.g. to carry in the dark. I'm satisfied that those Frenchmen went home convinced that they had met the stupidest young officer in the whole United States Navy!

We celebrated Christmas and New Year's in Falls Church, where Nick, barely one year old, was getting along famously. Then, after New Year's I was told one day by the personnel officer of the bureau that orders had been received from CinCPac, Commander in Chief, Pacific, for me to be detached from the bureau and report for duty to him at Pearl Harbor. I was absolutely delighted and saw this as my chance to escape the world of Washington and work my way to sea. Peggy took the news extremely well and decided that she would like to move Nick to California for as long as I would be away. With both my mother and my uncle, Grant Mitchell, living in Sherman Oaks, this seemed a good idea and we began making plans accordingly.

First there was the matter of my relief at BuAer. Personnel made it clear that there was no one in the bureau whom they could spare to take the job and that, because of its peculiar nature, they were going to leave it up to me to find my own relief. After thinking it over, I came to the conclusion that the best source for prospects might be the Air Combat Intelligence School at Quonset Point in Rhode Island. Personnel was cooperative and furnished me with the civilian backgrounds of all the reserve officers in the current class at Quonset. This class was due to graduate soon, and I burned the midnight oil going through their records. Among them was the curriculum vitae of Lieutenant Herbert Harlan, who was in his early thirties, was a professional writer, and had contributed articles to some aviation magazines. This seemed a very good background when one added to it the training in naval aircraft he would have received at ACI School. Of the entire class, he seemed the most appropriate, and I recommended to Personnel that he be ordered to the bureau as my relief immediately on graduation from Quonset. Personnel concurred and arranged his orders.

When Herb arrived on the scene, he was appalled at the nature of the job and was genuinely worried that he didn't know enough about naval aviation to fill it. I tried to reassure him that he could learn, and for two weeks kept him with me in what the Navy called "makee-learnee" status. We had a chance to go over the various facets of the job carefully and I came to feel that he could handle the routine part of it. Other assignments were unpredictable, as I found out on my own tour of duty, and since there was no way to prepare for them, I told Herb he would just have to take his chances, and call for help when he needed it. The civilian employees of the office provided continuity, and if he relied on Mrs. Marshall, he could always find out to whom he should go for information on a specific assignment. With great reluctance and much trepidation, Herb relieved me on 20 January 1943.

During those last few weeks, I was particularly busy on plans to convert the Bureau of Aeronautics newsletter into a greatly expanded and much more elaborate publication, *Naval Aviation News*. The concept for this transformation originated with Min Miller, and I supported it enthusiastically. Among other things, it involved making it a monthly magazine, printing it on coated paper to allow better use of photographs, and expanding and modernizing its format. After I left I believe that Training Literature took over its editorship, and it became such a fixture throughout the Navy that it is still being published.

My mother and uncle John, Grant Mitchell, agreed to look for a place that we could rent in the San Fernando Valley near Los Angeles. My orders to CinCPac staff gave us a few days before I had to report to San Francisco for transportation, so we packed up and headed for Los Angeles.

IV
Serving on Admiral Towers' and Admiral Nimitz' Staffs

With Nick in a market basket, we boarded an American Airlines DC-3 for what proved to be one of the roughest flights I have ever taken. At night over Arizona, the air became so turbulent that most of the passengers were violently ill, and I held on to the market basket for dear life as everything in the airplane not tied down went up to the roof of the cabin in the downdrafts. The DC-3s were not pressurized and consequently seldom flew above eight thousand feet: the only supply of oxygen was in a portable bottle which the stewardess might or might not have on board. Finally, in the middle of the night the pilot had had enough, and took us down through the turbulence to land at Tucson. The airline found rooms for us in the city, and we eventually got to bed about three o'clock in the morning. No sooner had we got to sleep than the telephone rang and we were told that the weather had cleared, the flight would go on, and we were to return to the airport!

Back we went to complete the flight to Los Angeles and a long ride from the airport to Sherman Oaks. With five days' leave before reporting to San Francisco, we enjoyed every minute of it and, thanks to Uncle John, solved the housing problem at the same time. Through him we were introduced to Mrs. Clive, the widow of a British actor who had recently died. She owned a lovely small house on the side of a hill overlooking the San Fernando Valley and was looking for someone to live in it and watch over it for an extended period while she was traveling abroad. We rented it for a very modest amount and I left with the satisfaction of knowing that Peggy and Nick would have a very nice place to live only a few minutes' drive from the house in which my mother and uncle were living. It was an ideal arrangement.

On 30 January, I reported to the headquarters of the Twelfth Naval District in San Francisco for transportation to Pearl Harbor. The city was an absolute beehive of wartime activity. At the District office, I was told that I would be leaving by ship rather than by air, and no one could say for certain on what ship or on what day. I was given a telephone number which I was to call every morning at eight o'clock and was to have my bags packed and be ready to leave immediately. If I was to go that day, the person answering my call would tell me what ship and pier to report to; otherwise I had no further responsibilities until eight o'clock the next morning. There were no Government quarters available and hotel space was almost nonexistent, but some kind soul told me that the University Club, across the street from the Fairmont Hotel, sometimes had rooms they would rent to transient officers. So up the hill I went, was courteously received at the University Club, assigned a very comfortable room, and given the run of the club for the duration of my stay. The charge for all this was much less than a hotel would have cost, and I have always had a warm spot in my heart for the club and its hospitality.

The uncertainty of not knowing from one day to the next when I would leave began to get wearing, particularly since I would not want to have failed to celebrate my last night ashore in the mainland United States. As one of the other officers staying at the club in the same condition of readiness said to me, "Every night is New Year's Eve, and by the time the ship is ready, I sure will be!"

After a week of waiting, my eight o'clock call turned positive, and I hurried down to the pier to report on board the USS *Kenmore*. She was a combination cargo and personnel transport and was carrying a mixed bag of perhaps fifty officers and Army civilian employees to their next assignments. As passengers, we had no shipboard duties, and after a pleasant but slow nine days of zigzagging, we arrived at Pearl Harbor on 16 February.

We were offloaded early in the morning and I was directed to the flagship of the Commander in Chief, Pacific, the USS *Pennsylvania*. She was one of the battleships that were sunk on 7 December and was resting in a huge dry dock along one side of the harbor. When I reported, I learned from the yeoman in the Personnel Office that all the other business of the command was conducted at CinCPac headquarters on Makalapa Heights, a nearby hillside overlooking the harbor. After logging in, I made my way to Makalapa and got my first view of the long, low cement building which Admiral Chester W.

Nimitz had fitted out as the command center of the Pacific Fleet. My first impression was that it certainly looked bombproof: tucked into the side of the hill, its sides and roof protected by very thick concrete, it had the appearance of a modern blockhouse. A little way up the hill from the building, I could see attractive-looking houses which obviously were where the officers were quartered, and the whole area was landscaped in the beautiful manner of Hawaiian residences throughout the islands. After inquiring of a sentry, I found the duty officer, Captain A.J. Wiltse, and reported for duty.

Captain Wiltse explained to me that although I was assigned to CinCPac staff, I was ordered for temporary duty immediately to ComAirPac staff and should lose no time in presenting myself to its headquarters on Ford Island, in the middle of Pearl Harbor. Accordingly, I found my way to a small-boat landing and chugged across the harbor.

Ford Island was the site of the Naval Air Station, which had been severely bombed and strafed by the Japanese; the piers around its perimeter berthed many of the battleships that were sunk or damaged on 7 December. Now, fourteen months later, the hulks of the *Arizona, Oklahoma,* and *Utah* were still clearly visible, but the hangars of the air station had long since been rebuilt and greatly expanded, and the runways were, of course, again operational.

ComAirPac headquarters was on the waterfront near the boat landing. When I reported for duty, I was promptly logged in and assigned quarters in the Bachelor Officers' Quarters (BOQ) overlooking the upturned bottom of the *Oklahoma*. After only a short wait, I was ushered into Admiral Towers' office and thanked him profusely for having sent for me. With a twinkle in his steel-blue eyes, he said, "Well, now that you're here, what are we going to do with you?" I repeated my hope that eventually I would serve in a carrier and said I thought my background as a civilian and in the Bureau of Aeronautics made me suitable for work in Air Combat Intelligence. The admiral was noncommittal, and simply indicated that I should report for duty at 0800 the next morning.

I did not know it at the time, but it was ironic that I expressed such interest in carrier duty, for that was the one thing Admiral Towers wanted for himself and had not been able to get. During his days as a captain, he commanded the *Saratoga* and gained a great reputation both as a ship handler and as one of the Navy's leading experts on carrier tactics. Unfortunately, his old difficulties with Admiral King stood in the way of his taking command of the carriers at sea. King reserved to himself the right of approving or disapproving the assignment of flag officers, and he was simply not willing to place the carriers in Jack Towers' hands. Although he had approved of Towers' promotion to vice admiral, King wanted him ashore and in an administrative and logistical job. As ComAirPac, Towers had vast logistical responsibilities for acquiring and distributing the aircraft, spares, and ammunition that the carriers in the Pacific Fleet needed, but he was precluded for a long, long time from taking any part in operational plans, let alone commanding at sea. King's attitude rubbed off on Admiral Nimitz, and it was not until later in the war that Towers, through his own persistence and expertise in carrier strategy and tactics, gained Nimitz' confidence, made his way into the top operational councils, and eventually became Nimitz' Deputy CinCPac and, belatedly, commander of the fast carrier task force.

For the next two months, I was in the awkward position of doing odd jobs first on ComAirPac staff and then back at CinCPac. While it was frustrating not to have a clearly defined assignment, I did meet and work with a number of very interesting people. Towers had assembled a congenial and able staff, led by Captain Forrest P. Sherman, who was considered to be one of the most brilliant thinkers in naval aviation despite having been in command of the carrier *Wasp* when she was sunk near Guadalcanal. With him at the time and now with him again on Ford Island was Lieutenant G. Willing Pepper in charge of Air Combat Intelligence. Commander David Ingalls, a well-known Cleveland industrialist and former assistant secretary of the Navy was busy organizing the Naval Air Transport Service, and at one time offered me a "permanent" job with it. Among the lieutenants on the staff were three Long Islanders from prominent families, Lee Loomis, August Belmont, and Oakleigh Thorne, the last serving as flag secretary.

After I had become well acquainted with most of the ACI officers, I was invited to join them in an almost unbelievable deal they had struck with the owners of the Halekulani Hotel on Waikiki Beach. For approximately thirty dollars a month apiece, seven officers in the group had rented a beautiful cottage under the palms on the grounds of the Halekulani, directly on the beach and adjacent to the Royal Hawaiian Hotel. Every officer on ComAirPac staff received one full day off per week and, by arranging the watch schedule accordingly, each of the seven officers took over the cottage for one twenty-four-hour period. It was a fantastic arrangement, and all the more fun for me when I discovered that Joe Pulitzer had rented a cottage for himself and was living there while he and Frank Rounds were attached to Public Relations in the Hawaiian Naval District office. We had a great deal of fun together, including a memorable party Joe gave when Artie Shaw, the famous orchestra leader, arrived on Oahu to become the leader of the Navy band. "You'll like Artie," said Joe as I was helping him get ready for the party, "he's been everywhere and he's been married to just about everybody!"

Not so congenial was the group that ran the Naval Air Station at Ford Island, whose tenants, as it were, ComAirPac staff had become. The base was under the command of a mustang, Captain John F. Wegforth, who was reported to be extremely bad-tempered, and I was advised to give him a wide berth. Little did I suspect that I would be with him later on under circumstances in which there would be no place to hide.

Incredibly, the CinCPac staff had only one naval aviator. He was Commander Ralph Ofstie to whom Joy Hancock had given me a note of introduction, and whom she married at the end of the war after he had commanded a carrier as a captain and a carrier task group as a rear admiral.

During my stay on the staff, I came perilously near getting trapped into Public Relations. The officer in charge of Public Relations for Nimitz was Commander Waldo Drake, whose only apparent credentials for the job stemmed from having been a waterfront reporter for a California newspaper. He was in the middle of what I suppose most often is an adversarial relationship between a wartime commander and the press. He fancied himself as a tennis player, and I unwittingly got off to a bad start by beating him severely on the CinCPac courts.

His deputy was one Lieutenant O'Callahan, who through no fault of his own, was unpopular with the correspondents because he was in charge of censoring their copy, which he was required by Drake to do in a very uncompromising way. At one point, while I was preparing a long paper on the different types of aircraft then in the Pacific Fleet, my desk was in O'Callahan's office. Obviously at Drake's behest, Cal called me to an after-hours meeting. "It seems," he said, "that two jobs that you could fill have opened up. One is in Public Relations on Admiral William F. (Bull) Halsey's staff down in Noumea; and the other is at sea in the North Pacific as Admiral Kinkaid's public relations officer. These are both good jobs. Which one would you like?"

I was extremely fearful that Drake would get me ordered to one of these jobs, and I would be stuck in Public Relations for the rest of the war. To explain my fear without offending O'Callahan and Drake was another matter. Fortunately, I had rehearsed my plea to get on a carrier so many times that I now gave a bravura performance, and both officers let me off the hook. Shortly thereafter, Captain Wiltse sent for me and gave me orders to report for training at the Pacific Fleet Radar Center in order to qualify as a fighter director officer. He explained candidly that even if I qualified, I might not be ordered to carrier duty, but if I did particularly well at the center, my chances were very good because a lot of new carriers were under construction. At long last the way seemed open.

V
Learning Radar and Fighter Direction

The Fleet Radar Center was also on the island of Oahu, but a considerable distance from Pearl Harbor. It was housed at one end of a Marine facility, Camp Catlin, and surrounded by barbed wire. At that time it was a seedy collection of rather ramshackle buildings, all wooden and all badly in need of paint. Along with its much fancier counterpart at St. Simon's Island, Georgia, the school at Camp Catlin was charged with training officers in the shipboard use of both air and surface-search radar, then still so new that the whole subject was shrouded in secrecy.

Under the command of Commander Jack Griffin, USN, a naval aviator, and his deputy, Lieutenant Commander Landon K. Thorne, USNR, the center undertook to teach the tactical applications of radar at a time when they had only begun to revolutionize carrier warfare, gunnery, and ship-handling. It was a particularly exciting time for naval reserve officers to enter the field, because it was an area of great responsibility, and one for which most regular naval officers had not been trained.

In the First World War, aircraft moved at speeds little greater than one hundred miles an hour and carried such light armament that they were not very effective against anything but other aircraft. In World War II, both Axis and Allied aircraft moved up toward a speed range of 400 miles per hour, and carried such powerful armament in their cannons, rockets, bombs, and torpedoes that they were a menace to any military objective afloat, ashore, or aloft.

With the science of military aviation progressing fast, it was becoming increasingly important and increasingly difficult to intercept enemy aircraft before they reached their objectives. Defensive fighters in World War II could not rely on the tactic of waiting over their ships until a raid came within sight, and then initiating their attack. To defend their positions effectively, they had to have warning of the approach of the aircraft long before they came within sight; and then go out on the right course and at the right altitude to intercept them as far away as possible from their objective. Planes traveling at 300 miles per hour would cover the last fifty miles to their target in ten minutes. Therefore, in order to intercept planes fifty miles away, the defensive forces would have to have twenty minutes' warning and start their interception run when the raiders were one hundred miles away, assuming that both they and the enemy were traveling at 300 miles per hour. Hence, the urgent need for the earliest possible warning of enemy raids; and hence the secret development of radar.

On aircraft carriers at sea, responsibility for the early detection of enemy raids and for maneuvering the defensive fighters into the best possible intercept position was centered in the fighter director officer aboard ship and his team of radar operators and radar-data interpreters. Collectively, this team and the compartment of the ship in which it worked were known as radar plot, later rechristened the combat information center (CIC).

Life and death issues hung on split-second decisions made by the men in radar plot. Every fighter in the air, limited by the bounds of optical vision, was subject to the orders of the fighter director in the ship, who was assisted by his radar vision, until the moment of actual contact between the fighters and the enemy. All aircraft, enemy and friendly, appeared on the radar screen as tiny flashes of light called blips or pips. If a blip showed a certain coded signal, it was a "friendly." If not, it was a "bogey" until positively identified as an enemy, when it was a "bandit." Upon his own responsibility and upon his own interpretation of the data provided by these flickering pinpoints of light, the fighter director in charge controlled the course, altitude, and speed of the fighters circling high above his ship by radioing his orders to them. Operating in a dimly lit compartment, surrounded by his radar indicators, his plotting boards showing the positions of friendly and enemy forces, and his radio equipment with which to talk to his fighters, the fighter director fought his war. His domain, which extended from sea level to 25,000 feet of altitude and from his own ship's position to every segment of the sky within a hundred miles on every point of the compass, was limited only by the range of his radar. To the men in radar plot, an air battle forty or fifty miles away was as personal a problem as one directly overhead; and, although tension mounted as raids came closer, there was always tension in radar plot. Like hunters, they looked with intense concentration for the first flickering sign of the enemy, twenty-four hours a day, day in and day out. Just as did much more senior officers on the bridge, they had to calculate risks, make instant decisions, and take immediate action on information that was often incomplete and subject to a wide variety of interpretations. And like any men who carry the responsibility for many men's lives upon their shoulders, they operated under a continuous strain commensurate with the stakes for which they played.

Not all the activities of a carrier's radar plot, however, were concerned with the enemy. Sometimes a routine flight of the carrier's own planes got lost in bad weather over the open sea; it was the job of radar plot to find it and guide it home. Sometimes a pilot had to make a crash landing in the water many miles away from his ship; it was up to radar plot to know where he landed and to direct a rescue team to move instantly to the scene.

There were many problems too that dealt with surface-search radar as differentiated from air-search radar. At night and in periods of bad visibility, the ship kept her station by radar, often while zigzagging at high speed in close company with scores of other ships, a situation in which there was constant risk of collision. Any object on the sea within the range of surface radar became the responsibility of radar plot, whether it was a friendly ship keeping a rendezvous, an enemy submarine or other hostile force, or just a buoy marking the approach to a narrow channel.

My class at Camp Catlin consisted of approximately thirty naval and two Marine officers. As at Harvard, we were a diverse but compatible group. Starting us out with an indoctrination in the mysteries of the cathode-ray tube, the curriculum took us through the applications of air-search radar, with heavy emphasis on fighter direction and the interception of enemy raids; how to tell our own aircraft from the enemy's by means of the top-secret IFF (Identification Friend or Foe); and the use of surface-search radar for station-keeping at night and in foul weather, for intercepting or avoiding other surface ships, and for inshore navigation and gunnery. In the process, we learned the plotting of targets; the fighter director voice code for communicating with fighter aircraft in the air; how to estimate the altitude of an air target by the way it faded in and out on the radar (there were no altitude-determining radars in existence); and how to use the complicated communications equipment for transmitting and receiving information throughout a ship or a group of ships.

For practice, we took turns at the one radar set in operation at the school and rotated through all the manned positions during miniature war games inside a simulated radar plot. Finally, to get the feel of a radar-controlled intercept from the pilot's point of view, we flew in two antiquated JRFs from Ford Island, one simulating a plane making a raid on Camp Catlin, the other the intercepting plane. They were slow two-seater amphibians, with the pilot in the front seat following the orders received by radio from the student fighter director, and a student in the rear seat monitoring the orders and trying to spot the target plane visually. Slow as the planes were, the experience did give us a feeling for the enormity of the sky, the use of cloud cover to avoid visual detection, and the importance of an altitude advantage and of being up-sun when you were trying to intercept.

During the whole nine weeks of the course, I worked extremely hard, my incentive being to better my chances of drawing orders to a carrier at its conclusion. Nevertheless, there was fun to be had along the way.

My roommates were the two Marine officers, Lieutenants Bilava and Howard. We shared half a brick house in Naval Housing at Pearl Harbor, and commuted to Camp Catlin in a jeep that Howard had scrounged from some Marine buddies at Eva Plantation. The Marines took their work at the Fleet Radar Center very casually in the belief that they would never be ordered to any kind of radar duty. Consequently, there was considerable hilarity around the house, augmented by the presence of an eighteen-year-old Hawaiian girl whom Howard persuaded to come in and cook for us, and who tended to take frequent showers and wander around the house clad only in skin with a US Navy towel draped around her neck. Howard was a farm boy and his oft-cited ambition was to return to Iowa at the end of the war, sit on his front porch with plenty of bottles of cold beer, and watch his bulls mate with the cows in the pasture. I hope he made it!

Since we had Sundays off, I kept my membership in the Halekulani group, which gave me a chance to keep in touch with my ComAirPac friends and with Frank Rounds and Joe Pulitzer. At one point, we were in dire peril of losing our little slice of paradise. It seems that one of our senior members threw a party that got out of hand, and caused the neighbors to complain to the owners. We would have been able to handle the ensuing ruckus, had the host not subsequently been found

Learning Radar and Fighter Direction

staggering around the grounds in the moonlight and relieving himself on the fender of an automobile which, of course, proved to belong to the owner of the whole establishment. Our lease was summarily canceled, and only by a tour de force in diplomacy were we able to get it reinstated.

One Sunday I got in touch with my cousin Marcus Monsarrat, a lifelong resident of Oahu, and he invited me to come to his house for the seventieth birthday party of his uncle, Bonnie Monsarrat. I was delighted to do so, and found the old gentleman a goldmine of Hawaiian lore. He was the first Hawaii-born person to go to Harvard and "read the law." When he returned, Queen Lilliokulani made him her personal attorney, and there was not much about Hawaiian history that Uncle Bonnie did not know. I found him fascinating and enjoyed Marcus as well. His son Roger, whom I had never met, was in the army and scheduled for duty in New Guinea.

Ever since I first knew him in Washington, Joe Pulitzer had been trying to get himself out of Public Relations and into a seagoing billet. It now developed that he had been successful at long last, and had orders to a destroyer. To celebrate these great tidings, he rented an exceptionally beautiful house on the beach at Lanikai on the windward side of Oahu. Here was held a week-long party complete with a Hawaiian orchestra-in-residence and all the whisky our pooled rations could produce. The guest list included every friend Joe had made on the island, and whole platoons of Navy and Army nurses. At the end of my session as a guest at this party to end all parties, Joe asked me to sign his guest book. After doing so, I noticed a space headed "Remarks" and entered after my name, "If this be war, why fight for peace?" Three days later at the conclusion of one of our classes at the Radar Center, Commander Thorne stopped me and remarked, "Glad to see you're back fighting for peace again!"

There were other interesting and more serious interludes. Twice we were ordered for a few days at a time to serve in carriers on training exercises at sea. I drew the spanking new *Essex*, the lead ship of a whole new class of 27,000-ton carriers and the first to arrive in the Pacific. In her radar plot, I learned how much I still had to learn, and it was a sobering experience. Later I was ordered to the British carrier HMS *Victorious* practicing joint operations with the US Navy. It was a most interesting glimpse of the difference between the systems used by the two navies. Unlike the American carriers, the British had armored flight decks, and I particularly liked their amenities of a daily ration of rum for the crew and an open bar before dinner for the wardroom officers.

At one point during my tenure as a student, I was promoted to lieutenant and was relieved to pass the required physical examination. At another, I was flabbergasted to receive verbal orders from Thorne to report immediately to Assistant Secretary of the Navy Artemus L. Gates, who was on a flying visit to CinCPac. Out I went to Makalapa in a cloud of dust, only to find that the secretary and his party were over at ComAirPac on Ford Island. By coincidence, Admiral Towers was just leaving CinCPac after his daily morning conference with Admiral Nimitz, and he consented to give me a ride over to the island in his barge, probably the fanciest form of transportation I have ever enjoyed. When I arrived, it developed that Mr. Gates had sent for me because he wanted his traveling companion, Samuel W. Meek, to have a chance to talk to me privately. This struck me as being strange indeed. Sam Meek was the vice president of J. Walter Thompson Company and in charge of its international operations. My only contact with him had been two brief interviews when he was looking for a young advertising man to go out to Calcutta to open a new office for JWT. I was intrigued at the time, but terminated further discussion when I discovered that Meek thought an appropriate salary might be fifty dollars a week. Now it seemed that he was a contemporary of Gates at Yale and was advising him on public relations matters, those of the Navy being in quite a mess. He was a contemporary also of my friends Bill Forbes, Rud Platt, and Deac Lyman, and someone, probably Deac, had told him that if he got out to Hawaii he ought to talk to me. Sam was not a very open person, and played his cards very close to his chest. I was not about to jeopardize my chance for sea duty by showing any wisdom about public relations, so I decided to answer his questions and volunteer little. He asked a good many questions, mainly about the complaints of the correspondents, and the internecine warfare between the black-shoe Navy, the battleship advocates and the brown-shoe Navy (the men in aviation). I answered as well as I could without putting my head in a noose, and made it very clear in the process that I was in training for carrier duty and had no interest in being assigned to PR work. We parted and I was glad to get back to Camp Catlin unscathed.

During the battle for Guadalcanal, Marine and Navy fighters and dive bombers based at Henderson Field were severely handicapped by lack of accurate radar information and fighter direction. Coast watchers behind enemy lines on the various islands to the north did a heroic job of spotting and reporting by clandestine radio all the enemy air and ship movements they could see; but there were severe limitations to that system. By June of 1943, the new *Fletcher*-class destroyers were coming into the fleet with SC air-search radar as well as the conventional SG surface-search radar. Commander, Destroyers, Pacific, ComDesPac, was headquartered at Pearl Harbor and, in conversations between him and Commander Griffin at the Pacific Fleet Radar Center, the decision was made to experiment with installing some kind of fighter director capability in the destroyers so that they might be able to control fighters available to them from other shore bases as the war progressed. To make a start in this direction, Commander Griffin agreed to lend DesPac two officer students who would install some kind of fighter direction setup in two destroyers and then operate in a carrier task group on training exercises to see if the system could be made to work.

After only a little more than a month of training, I was one of the two officers selected to plow this interesting new ground. During the second week in June, I was ordered to the USS *Bennett* a brand new, 2,100-ton destroyer of the *Fletcher* class, still the most beautiful ship I have ever seen.

The captain and the executive officer gave me full cooperation and I had virtually a free hand to improvise whatever system I thought might work.

After studying the layout of the ship, I decided to set up a plotting board in the navigator's compartment just aft of the bridge, and install a sound-powered telephone to the radar operators and to the bridge, along with a VHF radio receiver and transmitter so that I could communicate with the fighter director in the carrier. I then needed to train a bridge talker and a plotter and, when that was done, we would have the bare essentials in place. We had no time to meet with the carrier personnel in advance, and I couldn't even compare notes with my opposite number from the school, whose destroyer was berthed across the harbor. The only clues we had about the task group's operating plan were that, during the course of the exercises, land-based aircraft would be sent out to run a search and simulated attack on the task group, and that, at some point, the carrier's fighter director would turn over to the destroyers the job of intercepting a "raid" with the carrier's combat air patrol (CAP).

After three or four days of preparation, we put to sea in the midst of a hard blow. Because of their very narrow beam and light displacement, destroyers are notorious for rolling and pitching, and a sailor needs a certain kind of stomach to be able to function. I was billeted in a compartment to myself, and found that the best way to put on my trousers in the morning was to sit on the deck with my back wedged into a corner and wait for the ship to roll my way. When she did, I put both feet up in the air and pulled mightily on my pants legs until she began to roll the other way, when I could roll with her and come up standing with my pants on.

The area in which we operated was only some fifty miles away from Oahu, making it difficult to sort out the simulated raids from all the heavy island air traffic, all of course showing IFF. Because the carrier herself was flying antisubmarine patrols and conducting strafing practice on towed sleds, the air around us, as seen on the radar screen, seemed to this neophyte like one huge cauldron of unrelated pips.

We were given only one chance to try an interception. When the moment came and the carrier turned over one four-plane division of the combat air patrol to me, I was not certain which of the many airplanes between us and Oahu was the "raid." Consequently, I waited too long to vector out the CAP and, by the time it sighted the incoming raiders, it had far too little time to make a good intercept. We certainly could not claim to have given a convincing demonstration of the principle. Nevertheless I was satisfied that, given time to work out the problems, fighter direction from destroyers *was* feasible, and I so stated in my report when we returned to Pearl Harbor. Meanwhile, I had learned three lessons that were invaluable to me later on. One was to stay away from the muzzle of the 5-inch, 38 calibre antiaircraft gun. When this weapon was fired, it gave off an ear-splitting crack, and I had made the mistake of standing on the wing of the bridge far too close to the muzzle of one that was firing at a towed sleeve. For an hour I couldn't hear anything, and to this day my left ear misses a whole range of tones. The second lesson was that when low-flying torpedo planes come at you, surface-search radar can help you more than air-search radar during the last ten miles of their approach. Finally, I learned that some system should be worked out to cut down the interval between the time when the radar operator detects a raid and the relaying of orders to the combat air patrol. On balance, I may have learned more from the exercise than did DesPac.

As we approached the end of June, all of us at the Fighter Director School became increasingly anxious about the kind of assignment we might draw next. I knew that I had done well at the school, and felt

Learning Radar and Fighter Direction

reasonably confident that I would be assigned to carrier duty where the principal action was to be found for fighter directors. We all knew that many new *Essex*-class carriers were in various stages of construction, as were an even greater number of escort carriers, smaller and slower but nonetheless aircraft carriers.

VI
Carrier Duty at Last!

When our course was finished at the end of the month, our orders were issued from Washington. Mine directed me to proceed to Philadelphia, report to the precommissioning detail of the CVL USS *Langley* at the New York Shipbuilding yard in Camden, and then report for duty on board in connection with fighter direction. I was delighted and at the same time puzzled. I had never heard of a CVL and wondered what kind of carrier it was. Classmates Nick Stefan and Linus (Linie) Smith had received identical orders and they were as puzzled as I. Finally, we went to Lannie Thorne and asked him if he could shed any light. Lannie explained to us that CVLs were a new class of carrier, smaller than the *Essex* but just as fast, and considerably larger and faster than the CVEs. There were to be nine of them and all were being built in Camden on hulls that had been laid down as cruisers. With the urgent need for more carriers, the Navy had converted them on paper to carriers, and New York Shipbuilding was building them to new specifications. They were to be known as *Independence*-class carriers after the lead ship, which had not yet been completed.

I was detached from the Radar Center as a newly qualified fighter director officer on 3 July, and from CinCPac staff at the same time. After making my farewells at CinCPac and ComAirPac, I reported to the transportation officer to see about getting to the mainland. My orders called for "first available" transportation, including air, but the officer was not sanguine about my getting a place in a plane very soon. As had been the case in San Francisco, I was to check in every morning and await my turn. After several days of this, I finally asked the officer if there were any ships leaving for the West Coast. After some hesitation, he said, "Yes. There's a Panamanian freighter leaving for

San Francisco on 9 July. It has a Navy gun crew on board, and if you want to work your way back, I could get you assigned to the gun crew." "Great," I replied, "let's get it going."

This is how I came to report aboard the SS *Musa* in Honolulu. She had an English captain, a Scottish chief engineer, a Panamanian crew and a US Navy gun crew. Her one gun was a three-incher for firing at surfaced submarines, and she made her way without benefit of escort. For five uneventful days I stood watches with the gun crew. My sharpest recollection of the passage has to do with the *Musa*'s coffee: it was almost pure chicory, which apparently is the way the Panamanians like it.

Arriving in San Francisco on 14 July, I made tracks for Sherman Oaks for five days' leave and a joyful reunion with Peggy, Nick, my mother, and Uncle John. We agreed that I would see how far the *Langley* was from completion, and then let Peggy know whether it would be worth her while to come back east and be with me before the ship put to sea. When I determined that my stay in Philadelphia would make it worthwhile, she and Nick joined me and we rented an apartment in Haverford, near the college.

More than three years later, on a bright October day in 1946, Secretary of the Navy James Forrestal stepped into the park across the street from his office in Washington; called forward three men from the United States Ship *Langley* from the group standing at attention near the Mount Suribachi Monument; presented them with a Navy Unit Commendation for their ship; and read the following citation:

"For outstanding heroism in action against enemy Japanese forces in the air, ashore and afloat in the Pacific war area from January 29, 1944, to May 11, 1945. Operating continuously in the most forward areas, the U.S.S. LANGLEY and her air groups struck crushing blows toward annihilating Japanese fighting power; they provided air cover for our amphibious forces; they fiercely countered the enemy's aerial attacks and destroyed his planes; and they inflicted terrific losses on the Japanese in fleet and merchant marine units sunk or damaged. Daring and dependable in combat, the LANGLEY with her gallant officers and men rendered loyal service in achieving the ultimate defeat of the Japanese Empire."

These were warm words, spoken by a man who meant them. But when I first laid eyes on the *Langley* in Camden on 27 July 1943, no one was describing her so kindly.

She was then caught in that unhappy stage of construction, betwixt launching and commissioning, when tempers run short, frustration runs high, and the whole job seems so hopelessly involved that completion is impossible.

Freshly returned from Pearl Harbor and leave in California, I reported that morning to the ship's precommissioning detail, and with two others on the same errand, went down to the pier to look her over for the first time. None of us knew really what to expect. We picked our way through the busy shipyard between piles of raw materials, over railroad sidings, under huge traveling cranes, and around a long row of sheds crowded with workmen of every craft. At length we came to the water's edge, where, still at a distance, we could see three ships berthed at narrow finger-like piers jutting out into the Delaware River. A workman told us which was the *Langley*, and we moved up close to her side. Air hoses, electrical cables, and power conduits seemed to be wound around almost every inch of her like some jungle parasite vine strangling a tree. From every opening in her side we could see the flames of welders' torches in bright contrast to the grey Camden sky overhead, and we could hear the noise of a hundred rivet guns joining in a vicious, stabbing tattoo.

"Brother, she's fouled up like a Chinese fire drill. They'll never finish this job till the war's been over a month," said one of my companions.

"Looks topheavy. Bet she'll roll like a bowling ball," said the other.

With the broad area of the *Essex*'s flight deck still fresh in my mind, I thought, "How could you ever operate fighters from that little platform?"

We went aboard to climb down ladders, crawl along catwalks, and walk through the passageways that connected the various main compartments of the ship. Everything was raw steel. Everywhere the air reeked of the pungent welders' flux or of acetylene gas from the cutters' torches. Nowhere could you hear your own voice above the noise of the rivet guns. We soon gave up even trying to find the compartments that interested us most, and returned to the ship's office in the yard, where we found the tons of blueprints that revealed every detail of the ship's construction and equipment.

She was 623 feet long, had a beam of 71½ feet, a displacement of 11,000 tons and a draft of 26 feet; her flight deck measured 573 by 109 feet. Her complement was 1,400 officers and enlisted men, and her flank speed 32 knots. She was to operate only 24 fighters and nine torpedo planes; no dive bombers at all. Her biggest gun was a 40-mm. antiaircraft, so she must be meant to operate in close company with better armed warships.

Thirty officers and nearly two hundred enlisted men had already reported to the precommissioning detail. Paperwork and organization were the order of the day. Endless lists of equipment had to be checked and rechecked. Each division in each department had to draw up watch bills, allocate duties in accordance with its share of the ship's authorized complement, and lay plans for the training of the men.

During the next few days, as we who were responsible for setting up the radar activities of the ship began to analyze our problems, it became all too plain that we were going to be a green crew starting virtually from scratch. Over strong coffee in the mornings and stronger whisky in the evenings, we got better acquainted and assessed our meager pool of experience in the practical application of radar to carrier operations.

Lieutenant Morris R. (Dutch) Doughty had been designated fighter director officer of the ship and, as such, was the senior officer in the radar group. The only regular Navy officer among us, Dutch was also the only naval aviator. His record as a pilot was brilliant, extending all the way from ferrying fighters to Malta from the deck of the *Wasp* to months of action in the Pacific, where he flew in the dive-bombing squadron of the *Wasp* until she was sunk; he then operated from Henderson Field in Guadalcanal in the days when the Marines had only a handful of planes with which to oppose the daily runs of the Tokyo Express. But splendid as his record was, it was a flying record. His only radar training was in England where he spent a few weeks at the British radar school in Yeovil, in between cruises in the *Wasp*.

Gus Rounsaville, a broker from Dallas and also a lieutenant, was next in seniority. He graduated from the Fighter Direction School at St. Simon's Island, and had served in radar plot in the *Bunker Hill* during her shakedown cruise. In the two or three months that Gus was in the *Bunker Hill*, she went out only on training exercises and neither she nor her air group fired a shot in anger; nevertheless, he had been a watchstander, and was therefore regarded by our little group as the voice of experience on many matters.

I was third in line and the only other senior grade lieutenant.

Except for Tom Sorber, our technical expert and a brilliant young electronic engineer from Philadelphia, the rest of us were foreign to radar by civilian background. The remainder of our officer complement consisted of Linie Smith, a varnish manufacturer from Chicago, and Nick Stefan, just out of college in California, both of whom were classmates of mine at Pearl Harbor; Tom Draine, a university instructor from Chicago; John Hauser, an Atlanta architect specializing in the design of Methodist churches; and one later arrival, Bob Meserve, a lawyer from Boston.

Not trained in radar, but a very useful adjunct to the Gunnery Department, was Ensign Harry Shepard, a graduate of the aircraft-recognition school at Ohio State University and an expert in the instant identification of our own and enemy types of aircraft. We referred to Harry as "the poor man's radar."

Collectively and individually, we nine were charged with the responsibility for supervising the installation of the ship's three radar sets. It was up to us to equip radar plot with the thousand and one items that would, or might conceivably, be needed for the practical interpretation of the information that our radars would provide, and to weld ourselves and our radar crews into an efficient and, above all, reliable team.

Underlying these responsibilities was a fourth and even deeper one, harder to define but as clearly recognized. Radar was a new science at this time, and those who understood it were few and far between. Unlike gunnery, navigation, engineering, communications, and the other special branches of seagoing operations, radar was not then taught at Annapolis; therefore, even the most recent graduates of the Academy knew practically nothing about it. This made it almost entirely a reserve officer's specialty, and at the same time deprived the reserves of the hard core of regular Navy experience on which they could usually draw for guidance. To some captains, radar seemed at first like just another complicating gadget in their ships. To those who had read or heard of apparently miraculous feats accomplished with its help, it seemed like a magic cure-all, a nostrum on which they could rely to extricate themselves from nearly any kind of trouble.

To the reserves, then, fell the delicate task of educating their regular superiors to the possibilities as well as to the limitations of radar, and of doing so in a way that would create confidence rather than animosity. The ideal curriculum at radar school would have included several semesters in Machiavellian diplomacy.

The nine officers who set out to solve these problems for the *Langley* recognized one fundamental fact from the beginning: that just as the success of the carrier depended to a considerable extent on the success of her radar plot, so the success of radar plot depended on the ability of the enlisted men who manned the radar sets. The sets might be in perfect operating condition—we had to rely on Tom Sorber to see that that was so—and our own training might in time make us capable of correctly interpreting the data we received from the sets; but unless the data were as accurate and as complete as was humanly possible, we could never even hope to reach the peak of usefulness and reliability.

Once this was agreed upon over steaming black coffee in the dingy "wardroom" of the Wellsbach Building in the Camden shipyard, we turned eagerly to the thick sheaf of papers that contained the lists of the ship's authorized complement. We liked what we found under "radar plot": an imposing list consisting of two chief radarmen, five radarmen first class, seven second class, and ten third class. This was an impressive array of talent, and in our innocence we waited eagerly for the first man to report aboard.

When none had come by the end of the week, some of us began to get a little anxious. By that time we had learned that our sister ship, the *Cabot*, was in the navy yard just across the river and almost ready to leave on her shakedown cruise. I wanted to inspect her radar plot and try to get a few words of wisdom from our opposite numbers. Accordingly, Tom Sorber and I got permission to visit her, and returned with the news that only three or four radarmen of *any* class had been assigned to her, and all her other operators had to be trained from scratch. This was the first of many lessons we learned concerning the fact that a wide gulf lay between the neatly mimeographed lists of "authorized" men, supplies and equipment and the men, supplies, and equipment that actually made their appearance.

Over a drink that night in Philadelphia, Tom, Linie Smith, and I pondered the dilemma. What was the use of a complement list if you didn't get the men to fill it? How were you supposed to know whether or not you were going to get *any* of the men who were authorized? What would you do if you didn't . . . put to sea without any radar, or just quietly cut your throat?

By the time our glasses had been refilled, we had agreed on one thing: that however bad the truth might be, it was to our advantage to know it as soon as possible. We parted with the understanding that we would take the problem to Dutch in the morning.

Dutch had less time to devote to radar problems than he would have liked. As V-3 division officer, he was responsible for the organization of air plot, aerology, and the photographic laboratory as well as radar plot; necessarily he had to delegate many responsibilities. Accordingly, when we told him our fears the next day, he left it up to the three of us to investigate further and take whatever steps we could. Like dogs on the scent, we headed across the river for the Personnel Office at the Philadelphia Navy Yard. There we found that several hundred of the *Langley*'s crew, most of them just out of boot camp, were already assembled and others were arriving daily. A few were being sent over to Camden, but only a few, as the quarters at the shipyard were relatively limited.

"Who has custody of their service records?" we asked a grizzled chief petty officer.

"The commander in the Personnel Officer's office," he replied; and then, as an afterthought, "But most of 'em haven't had much in the way of service to record."

Up we went to find the commander and to hold the first of what became a long series of meetings. The situation was "very ungood," as Tom described it later. The *Independence*, first of our class, had only a handful of radarmen aboard. The *Cowpens* had gone out with only five. The *Cabot* would soon be sailing, and could not expect any more than she already had. And the *Langley*?

"Well," said the commander, "If I were you I wouldn't expect more than five or six trained men. You'll just have to train the rest yourselves."

He said it in a way that implied you could cut out radarmen like cookies from a cake of dough, and we suspected that he had only the vaguest notion of what a radar set was or of how complicated it was to operate.

My next question to him brought forth an answer that was no more reassuring. "Assuming that we don't get anything like our authorized complement, who decides which other men we do get, and where do they come from?"

"Well," he replied, "I generally make that decision over here, and assign you the total number of men you need from the pool of seamen second class assembled here in the yard."

"Excuse me, Commander, but would you mind telling us just how you go about picking them?" I asked.

"Young man, I've got too much to do to go about hand-picking them if that's what you mean. I know my business. I used to be the personnel manager in a big insurance company before I ever got mixed up in this rat race. When the crew for a ship like yours comes in here, my men and I go through their service records and assign the petty officers to the divisions they're trained for. Then we take the seamen seconds and prorate the ones with some experience among all the divisions, and divide up all the rest, division by division, until the complement is complete. We can't do any more than that. I suggest that if you're not satisfied with the men you get, you try to trade 'em off with some of the other divisions for men you think you'd like better."

By the time the commander had got halfway through this statement, he had begun to sort through the pile of mail on his desk, and it was obvious that, as far as he was concerned, the sands of our time had just about run out. I decided to take the bull by the horns.

"Sir, you must be awfully busy with all these ships to look after, and each one with a whole batch of individual problems, and not enough men to go around. It must be a really tough job, and we certainly don't want to make it any tougher by asking for special treatment. At the same time, it's particularly important to us to get just the right men for this radar work in the *Langley*. I wonder if we could make a deal with you that might help us both. If you will give us access to the service records and the qualification cards of all the seamen assigned to the ship, we'd like to make our own selection of the ones assigned to radar plot. We'll promise to stay out of your hair in the process, and after all, it will be saving somebody on your staff some clerical work, even if it's only a little."

There was a perceptible silence, before he answered. Then: "I don't think you realize that picking men for specific jobs is a science in itself and calls for a long training and experience. Unless one of you men is a psychologist, I think you're wasting your time. But if you promise to stay the hell out of here, and promise not to tell anybody else on that ship what you're doing so I don't get every division officer in the *Langley* telling me how to run my business, I'll let you do it. Tell the chief in the outer office to let you look at the records, and now, if you please . . . ."

Tom, Linie, and I thanked him and got away as fast as we could while we were still ahead of the game. Stopping at the chief's desk on the way out, we explained briefly that we had the commander's permission to look at the records, and asked him to show us one as a sample of the kind of information they contained. He gave us a look that clearly indicated his belief that gold stripes could never take the place of hash marks, and picked one out of the index cabinet alongside his desk. On it was a considerable amount of compressed information, compiled no doubt from the reams of forms that we knew from experience had been filled out on every man who ever entered the naval service. In addition to name, age, birthplace, and other vital statistics, there was a key letter that gave a clue to the man's IQ, and a description of his civilian occupations and hobbies. We thanked the chief, found out that he had four hundred on file so far, and told him we'd be back the next day.

Twenty minutes later, at the bar in the officers' club at the navy yard, we took stock of the situation over martinis. What characteristics would tend to make a good radarman? Obviously, intelligence enough to grasp the importance of the job, and to master the intricate mechanism involved. The IQ key could give us a guide on that. Obviously too, the man's eyesight should be above minimum Navy standards, so that he could make the fine visual distinctions that would become his stock in trade. The card would give us a factual answer on that. Manual dexterity was important when it came to operating the maze of knobs and cranks that were integral parts of the gear. The card would help there by telling us what the man had done as a civilian and what his hobbies were. But how could you measure the intangibles? How could you tell whether a man had the patience to stand watch after watch staring at a little jumping trace of green light that flickered continuously before his eyes? Or whether he had the power of concentration? Or whether he would get claustrophobia living the better part of his working life in a darkened room about the size of a bedroom closet? Or whether he would get uncontrollably excited when he saw enemy aircraft appear on his set?

I think it was Linie Smith who made the suggestion. "I think it would be great," he said, "if we could get men who *want* to try this job."

Accustomed as we were after a year in the Navy to thinking in terms of orders rather than desires, this came as a sensational idea. It led us immediately to the thought that we might set up a preliminary screening: on the basis of the service record cards, we could sort out the most likely candidates, explain to them the nature of the work in as much detail as its secrecy would permit, then eliminate those who didn't respond favorably to the challenge.

"If BUPERS finds out about this, they'll have a hemorrhage and send us all to Mine Disposal," I said.

"They can't," said Linie, "unless that commander tells them, and besides we'll be at sea and they'll get our orders lost before we get back."

And so it was decided.

Carrier Duty at Last 51

We sorted through every one of the cards the chief had on file, and picked out eighty-five men, each of whom had some characteristic that made us think he might make a good radarman. The next problem was to get them together for our explanation of radar work, and for personal interview. If we sent out the request to see these men through normal channels at the barracks in the yard, questions were bound to be asked and the cat would be out of the bag; so we decided to take the chief in the Personnel Office into our confidence.

"It ain't Navy," he said, "but you've gone this far; I guess I can get 'em over here for you."

In my talk with the first group, I tried to explain radar and its importance to carrier operations as frankly as security regulations permitted. After telling the men the general purpose of the meeting and emphasizing that no one would be drafted for the job, I gave them the following idea of what it was all about:

"Radar is a word manufactured from the initials of 'Radio Detection and Ranging.' It is an electronic device for detecting the presence and exact location of surface ships and aircraft at much greater distances than you can see them with even the strongest binoculars. It works as well at night, or in heavy clouds, as it does in the daytime or with a cloudless sky. It also enables you to tell whether or not the ship or aircraft you detect is friendly or belongs to the enemy. By analyzing the data that radar provides, we can determine the course and speed of the object detected, and having determined that, we can determine the best course and speed for our own aircraft to intercept it before it can get close enough to attack us. The ship herself and the lives of all her crew will probably depend many times over on how quickly our own aircraft make these interceptions. The fighter director officer is responsible for directing the interceptions by using his radio to order the fighter planes to the right point at the right time. But unless the information the radar operator gives him is complete and accurate, he cannot do this job, and every man in the ship may pay for it.

"If you are chosen for this duty, you will work at operating the ship's three radar sets. They are probably the most complicated devices you've ever seen; they look like a cross between a large radio, a juke box, and a fortune teller's crystal ball. We can teach you to operate them, but don't get the idea that it's easy or glamorous. It's difficult and monotonous and it's no place for a man without patience. It's also secret and you won't be able to tell your family much about what you're doing.

"In addition to working on the sets, those of you who are chosen will work in radar plot. That's where the information that comes in on the various radars is recorded, analyzed, and acted upon. It's no place for a nervous man. The decisions made in radar plot in a matter of seconds after a piece of information comes in may decide whether a ship is sunk or stays afloat. Pilots' lives depend on them. Your own life may depend on them.

"If you undertake this work you will be undertaking one of the greatest responsibilities of any enlisted man in the ship. You can take a corresponding amount of pride in it. But because so much depends on this kind of work, no doping off can ever be tolerated. The first time you're caught soldiering on the job, you're in real trouble.

"Now if any of you men feel that you want to try it, stay here after this meeting so that Mr. Sorber, Mr. Smith, and I can talk to you individually. The rest may leave."

Twenty men of the first group of twenty-five stayed for further talks, and of this number we chose eight. In the succeeding seven to ten days, we selected twenty-two more. This provided us with six over our authorized complement of twenty-four, without allowing for the appearance of a rated radarman. After discussions with Dutch, we decided to have all thirty ordered to duty with V-3 Division, keep them all in radar work until their first training was completed, and then transfer the six least acceptable to other activities in the division.

Since our radars were not installed in the ship at the time, much less in operating order, the best thing to do seemed to be to send the men to the radar operators' training school in Norfolk. There, even though they would not be under our supervision, they would "learn by doing" and begin to get some practical, as opposed to theoretical, training.

One hurdle remained: getting the captain's permission. We had not yet had any contact with him except the pleasant formalities of reporting for duty, and we wondered how he would take the idea of sending thirty newly arrived men away so soon. With the blessing of the air officer and the executive officer, we called on him and explained what we wanted.

"Seems like a good idea. Go ahead," he said.

The first obstacles were over; we would put to sea with twenty-four men who had at least seen a radar set, whether or not BuPers sent us any. I think we were all pleased. But none of us could then foresee the dividends that later accrued from our method of selecting volunteers.

During the first few weeks after we had packed our thirty men off to radar school, time began to pass more quickly and things looked more hopeful. The ship herself slowly but steadily emerged from her cocoon of power lines and compressed air hoses. The mountains of paper work in the Wellsbach Building gradually dwindled down to mere hills, and more and more officers were reporting "on board" to help us attack them. And at the navy yard across the river, the pool of men assigned to the *Langley* grew to sizable proportions.

Eventually the day came when the ship was far enough along to move across the river and exchange her Camden Shipyard berth for a regular Navy berth in the yard. This transfer was accomplished by tugs, and while it was a far cry from actually being ready for sea, at least she was moving, and at least she was free from the first group of her builders. At her pier in the navy yard, she was promptly attacked by a new swarm of workmen intent on the thousands of details connected with fitting her with all her government-furnished equipment. Here she was to get her guns, her ammunition, her radar and radio equipment, her navigation equipment, her laundry, her galleys, and the hundreds of other items needed to make her a ship in which fourteen hundred men could live and fight.

Radar plot was still an empty compartment, about fifteen feet wide and twenty-five feet long, and many decisions had to be made as to the arrangement of its complicated equipment. The interpretation of radar information was still so new a science that no standard arrangement for radar plot had been prescribed by the Bureau of Ships. The bureau merely specified what types of radar, radio, and intercommunication gear were to be installed, and left the fighter director officer wide latitude in planning the layout for his individual ship.

Accordingly, we spent the first few days after our ship's transfer to the navy yard in visiting as many ships as possible to inspect their radar plots and benefit from their ideas. Gradually, a layout plan evolved in our minds, and with almost daily mental erasures and changes, we reduced it to paper and worked out its details with the superintendent of the workmen assigned to our compartments in the ship.

In general, our plan was to separate physically the two main functions of radar plot: the plotting and analysis of surface targets and the plotting, analysis, and provision for taking action on aircraft targets. At the end of the compartment farthest from the fighter director's position, we placed a chart table on which could be plotted our ship's position in relation to any land or any other surface ship detected by the SG surface-search radar. In the center of the compartment we put the main plot of all air targets reported by the SC and SK air-search radars; and, in accordance with the best of the radar plots we had seen, we asked that this main plot be a circular piece of Lucite, six feet in diameter, engraved with compass bearing lines every ten degrees and range circles to a distance of 140 miles, mounted vertically with provision for edge lighting around its entire circumference. Behind this transparent plot, our enlisted plotters were to mark with grease pencils all targets reported by the two radars, marking and writing backwards, so that their writing would read properly from the forward side of the plot. In front of this main plot we arranged two smaller horizontal plots, also of Lucite, on which we could concentrate only such information as pertained to individual raids that we were assigned to intercept, or individual surface problems that it was our duty to solve. In the center of one of these, we placed a ship's model attached to a gyro, which would enable us to tell at a glance the exact bearing of any approaching target relative to her heading at any given moment. Handy to all three of our plotting boards, we positioned our two remote indicators which would show us at the flick of a switch the exact image that appeared to the radar operator working on any one of the three sets; we then arranged the various radio and intercommunication sets with which we could talk to aircraft, other ships, our own radar operators, the bridge, the lookouts, and all the key positions above and below decks.

After agreeing on this basic arrangement, we began to add refinements to the layout according to ideas about which several of us had come to feel strongly. Dutch wanted air-conditioning, in case the ship went to the tropics. I wanted comfortable stools for the plotters who would have to stand long watches. Linie wanted a status board big enough to keep track of all aircraft in the air by their individual numbers. Tom Sorber was worried about keeping the compartment dark enough to allow the low lights of the repeater indicators to be followed, but light enough for us to see our way around.

During the next few weeks, each of us worked on his own particular part of the job, scouring the navy yard, pestering the supply officer, and scrounging around Philadelphia for items of equipment to make this one compartment of the ship ready.

For some time I had been working on the design of an entirely new scale ruler to increase the accuracy with which we could dead-reckon the course of the Grumman F6F fighter plane. With the help of

Lieutenant Filo Turner, the engineering officer of VF-32, I now had accurate data with which to make this invention possible. By calibrating the marks on the scale with the mile-scale of a standard Navy plotting chart, I designed four sets of graduations: one for each of the speeds we used in the fighter director code—Saunter, Liner, Buster, and Gate. The ends of the scale were rounded to represent the tightest ellipse that it was practical for a fighter to fly when it was ordered to orbit. After making a precise working drawing of this new instrument, I took it to Keuffel & Esser in Philadelphia and had them make a dozen Plexiglass scales, paying for them out of my own pocket.

When the air officer, Commander E.A. Hannegan, saw me working with the scale one day, he became interested and sent a description of it with a photograph to the Bureau of Aeronautics. Nearly a year later, I was pleased to see a circular letter from the bureau recommending the device to other carriers.

It fell to me to have the main plot made. "Go and see a man named Max Levy in North Philadelphia," suggested the yard supply officer. "He made the plots for the last three ships that went into commission. He doesn't know what they're for, and of course you can't tell him, but he does good work and doesn't ask questions."

I found Max in a small machine shop on a little street near the North Philadelphia station. Luckily, he had saved the working drawings of the plotting boards he had made before and, with a few simple alterations, they would fit our specifications.

"You get me the Lucite, and I'll make it for you," he said. "But some day when this war is over, I hope one of you guys will come back and tell me what the hell this is all about."

We learned later that he made plots for every carrier that left the yard and wondered if he ever found out to what use they were put. If he had known the hundreds of raids that were to be plotted on ours alone!

Day by day, all departments of the ship showed signs of progress. Before long we had enough of our equipment on board to be formally commissioned. On 31 August the ship's officers and enlisted men fell in at flight deck parade, Captain Wallace M. (Gotch) Dillon read his orders, the commission pennant and the ensign were hoisted to the breeze, the first watch was set by the navigator, and we were, at least theoretically, in business.

Our thirty hand-picked men returned from radar school, each graded by his instructors according to the aptitude he had shown and the progress he had made. We conducted personal interviews with each, and all but three expressed the desire to continue in radar work. We promptly released these three to other assignments within the division, and reassigned the three who had made the poorest showing. This left us with our complete complement of twenty-four. Only one other radar operator was assigned to us by the bureau, so scarce were personnel in this category.

A few days after the commissioning ceremony, we were ready for our acceptance trials, and left the dock under our own power for the first time. Even though we were only cruising up and down the narrow channel of the Delaware River, this event was the most important milestone yet. In the crew's quarters as in the wardroom, it raised everyone's spirits, lessened the grumblers' grumbling, and underscored for all hands the fact that this was, after all, a ship designed to do a ship's work.

After the acceptance trials and before leaving on our shakedown cruise, we instigated an intensive course of training for our men. Tom Sorber drilled them daily in operating the sets, using civilian airliners approaching Philadelphia as air targets and shipping on the Delaware as surface targets. John Hauser, the architect, taught them plotting, including the intricacies of writing backwards on the vertical transparent plot. I taught them the theory of intercepts, how to work simple maneuvering-board problems and the basics of aircraft performance and tactics. Linie and Gus drilled them in communications, and Tom Draine held daily lectures on their general duties around the ship. We tried in every way possible to teach them not only their own jobs, but the fundamentals of their officers' jobs as well, so that they could get the feeling of what the unit as a whole intended to do.

One major job remained: to get acquainted with the fighter squadron that was to be assigned to the ship, and to practice intercepts with them, both for their own and our experience. To accomplish this, we left Tom Sorber in charge of finishing up the radar installations in the ship, and all the other radar officers went up to Quonset, Rhode Island, where the fighter squadron was based preparatory to joining the ship.

Working with the radar installation at Beavertail Point, we spent a week running day and night intercepts, half the squadron acting as the enemy and half as our own intercepting fighters. This daily exercise gave Dutch his first chance to see his officers at work on actual intercept problems, and to make his evaluation of where each of us would best fit into the scheme of operations in the ship. All of us were rusty from lack of practice, and none of us was familiar with the gear at Beavertail, but the week accomplished its purpose and, when the squadron eventually came on board, we were not strangers to each other.

On our return to the ship, it was apparent that we would soon be ready to leave on our shakedown.

"How long do you think we'll be gone?" I asked Dutch.

"Six or eight weeks, I imagine."

"Long cruise."

"The *next* one's the long one," he said with a smile.

58   Angel on the Yardarm

USS *Langley* at cruising speed. Photo from National Archives.

VII
Shaking Down the *Langley*

Two new *Fletcher*-class destroyers accompanied us on our shakedown cruise to the Gulf of Paria, between Trinidad and the Venezuelan coast. We steamed down the Delaware River, picked up speed for the transit of Delaware Bay, and, as we rounded Cape Henlopen for the open sea, the destroyers began to pitch and roll in the traditional manner of the "tin can Navy." The seamen in the *Langley* looked out across the water at them with the curiosity born of tall tales heard in boot camp.

"Glad I'm not in one of them," said one sailor.

"You may live to change your mind," replied his mate.

"Or not live at all," rejoined a lugubrious third.

Down in radar plot we shook down quickly. It was a relief to get to sea, away from the confusion of heavy land indications on the radar screens. With two officers and six enlisted men on duty in each four-hour watch, we found we could comfortably handle the routine work, while at general quarters each day, we were able through experience to eliminate more and more of the rough spots in our training for the real thing.

By the time we were approaching Norfolk, where we were to make a brief call, all hands in radar plot were beginning to feel salty and more than a little pleased at the lack of troubles so far. This complacent frame of mind received a rude shock late in the afternoon, when the captain called Dutch to the bridge, put his pencil on a chart of the Chesapeake, and said, "This is where I want to anchor tonight. The weather's getting thicker all the time, and it'll be dark by the time we enter the bay, so I want you to guide the officer of the deck into the anchorage by radar."

Dutch brought the chart down from the bridge and together we pored over it. The main channel led into Chesapeake Bay in a westerly direction between Cape Henry to the south and Cape Charles to the north. About halfway in, an auxiliary channel branched off to the north toward the place where the captain had placed his pencil mark. None of us knew whether our SG radar would even pick up the small channel buoys bobbing in the turbulent sea.

At dusk, with considerable relief, we picked up the entrance buoys between the capes before they could be seen from the bridge and, as the ship steamed up the main channel, we plotted our progress on the chart table and passed ranges and bearings on each pair of buoys up to the bridge every minute. The next thing we needed to do was to warn the officer of the deck of his approach to the junction of the auxiliary channel, and we made a discreetly couched suggestion via our "talker" on the bridge that he slow down from his present speed of 15 knots.

"Captain wants to come in fast," was the curt reply down the "squawk box" suspended from the overhead at the fighter director's seat. At this point, Tom Sorber and I began to perspire, although the day was far from warm.

Together, we relayed the word to Dutch that the turnoff buoy would be the third next in line on the starboard hand, and hoped devoutly that what we thought was the buoy did not turn out to be a small boat anchored on the shoal at the side of the channel. It wasn't. We made the turn, and saw on our repeater screen pair after pair of buoys leading us north to the vicinity of our anchorage. As we passed through the last pair, the talker at my elbow said, "Bridge requests course and suggested speed to the anchorage."

Tom had worked out the course while I was busy plotting positions, and the slower we went the better we would be pleased, so I was able to reply instantly, "Course Zero Three Five True, suggested speed 5 knots."

Our gyrocompass repeater swung quickly to the course indicated and the log began to drop slowly from 15 down to 10 and finally to 5. With no more buoys ahead to guide us, we began to take cross bearings on two points of land to fix our position, and reported ranges and bearing to the anchorage. As the range decreased, we suggested slower and slower speeds, which our log told us were being followed on the bridge. None of us knew how much distance to allow for the ship to carry her way after her engines had been stopped, and I undertook to estimate this on no more scientific a basis than my experience under sail in small boats. At the suggestion that all engines be stopped, our talker on the bridge reported that the officer of the deck had ordered them

Shaking Down the *Langley*   61

stopped, and a few minutes later, our triangulation showed us it was time to drop anchor.

I sent up the word, "You're on the range. Suggest drop anchor."

In a matter of seconds we heard the rattle of the anchor chain. After allowing a few minutes for the captain to go down to his sea cabin in the island, two of us went up to the bridge to see what we could see. Through the port and starboard alidades, we took visual bearings on lights, which we were sure the navigator had been doing all during our approach, and they seemed to check almost exactly with our own.

The next morning, we climbed to the bridge again and took bearings on the points of land that were then clearly visible. Pleased with the result of our first really useful job, we went below for breakfast. On the way into the wardroom, we passed the navigator.

"Not too bad," he said, "but you fellows were about three hundred yards off last night, according to the readings I took this morning. Don't let it happen again."

It may have been a coincidence, or it may not; but during the remaining twelve months that the navigator was in the ship, none of us in radar plot ever warmed up to the man.

After a short call at Norfolk, we left for Trinidad and the Gulf of Paria with Air Group 32, twenty-four Grumman Hellcat fighters, and nine Grumman Avenger torpedo planes on board. The real work of the shakedown then began, with a heavy daily schedule of air operations, gunnery drills, and fighter direction exercises.

One of our main concerns in radar plot at the time was to compile a mass of accurate radar data from which we could estimate the altitude of any given aircraft target that we detected on the screen. No radar set yet devised would give a direct reading of the altitude of targets, and unless we had the basis for a reliable estimate, the fighters that we sent out to intercept enemy raids might be so far off in altitude as to miss the raid entirely or be at a severe altitude disadvantage. Theoretically, altitude could be estimated within one thousand feet or so by the range at which a target was first detected; the longer the range of initial contact, the higher being the indicated altitude of the aircraft. Also theoretically, it could be estimated by the intervals at which the radar indication of the target periodically faded from view on the screen during the target's approach toward the ship. But to translate the theoretical solutions by either method into solutions that would be dependable in actual practice required carefully controlled flights at every thousand feet of altitude from one thousand to twenty-five thousand; with each flight extending to a range beyond the point of initial pickup of each radar set.

This was a tedious task for the pilots, and a time-consuming one for the air officer to include in his already crowded schedule of air exercises. Before leaving Philadelphia, we carefully explained the need for it to the air officer, but on the way down to Trinidad only a few such runs were scheduled in the air plan of the day. We, therefore, raised the subject again, and were assured that time would be allocated to it during our stay in the Gulf of Paria.

Steaming southward at 18 knots, we soon crossed the Gulf Stream, skirted the Bahamas, and set our course for Mona Passage, between the eastern tip of the Dominican Republic and the western tip of Puerto Rico. A day short of the passage we picked up our first radar contact that resembled a submarine. It appeared shortly after midnight, a small, clear indication well astern of our little group. We sent its range and bearing up the squawk box to the bridge, and marked it carefully on our surface plot. The radar operator gave us a fresh range and bearing every minute and, with our third plot, we were able to solve the maneuvering-board problem, which gave us the object's course and speed. If it was a submarine and our calculations were correct, it was too far astern to come within torpedo range and the officer of the deck ordered us merely to watch it closely and report every two minutes. Three plots later, while it should still have been well within radar range, there was no sign of it and it never reappeared. Presumably it had submerged.

The next night we picked up our first radar indications of land and, as we drew near to Mona Passage, we received strong pips from the mountains on Hispaniola. The transit of the passage was done by radar in the early hours of the morning, and the first light of a beautiful dawn gave us visual verification of our radar position, Mona Island abeam to starboard and the rugged shore of Puerto Rico to port. After the strain we had gone through trying to detect the buoys at the mouth of Chesapeake Bay, this was child's play, and we all thoroughly enjoyed it. The following few days took us down past the Leeward Islands, past Martinique and the Windwards, and into our anchorage in the Gulf of Paria at Port of Spain, Trinidad.

Our next two weeks were filled with more training exercises than ever, but the plan called for spending nearly every night at anchor, with liberty ashore for one watch at a time. We had been told fabulous tales of the wonders of the famed Macqueripe Club, just outside Port of

Spain, and on our first night in port the officers of the starboard watch took to the whaleboats en masse to discover whether the tales were true. Being a member of the port watch, I had the eight-to-twelve watch as officer of the deck, and just before I was relieved at midnight the first boatload came straggling back. While they were still a hundred yards from the gangway, it was evident that they had had an active evening. By the time they climbed up the ladder and stepped into the light on the hangar deck, it was apparent that it had also been an athletic evening. Several had braved the swimming pool fully clothed. The torn shirts of others bore witness to the extent of "an argument with some bastards on the beach." Almost everyone had carried the pitcher to the well more than once too often.

At dawn the next morning we put out into the gulf, had a strenuous day of flight and gunnery operations, and returned to receive a blinker message just before the port watch went ashore: "Effective immediately, the Macqueripe Club is out, repeat out, of bounds to officers from the *Langley*."

I'm told that it was great while it lasted, and except for the scenery, no other feature of Trinidad seemed to live up to the Elysian tales that were henceforth heard about Macqueripe by the port watch.

The most interesting exercises we conducted in the gulf were night-qualification landings for the air group. These were held on cloudless nights, under a full tropical moon bright enough to reveal the mountains of Venezuela, thirty miles across the water. One by one, each member of the air group followed the group commander in, with nothing but the moon, the tiny glow lights set into the deck, and the landing signal officer's paddles to guide him. It was a beautiful sight: first, the red dot of the airplane's port wing-tip light streaking past the starboard side of the ship; then the plane dipping as the pilot crossed our bow and headed on the opposite course; then dipping again as it banked and turned to come "up the groove" over our phosphorescent wake; finally the flash of paddles, the solid smack of wheels on deck, and the welcome sound of the barrier's fall to clear the way for the pilot to taxi up the deck. No evolution so clearly illustrates the precision training of Navy pilots as does the thrilling commonplace of night landings on a carrier at sea.

Two tragedies and one disappointment marked our stay.

On an early-morning launch, we lost our first pilot and crew in the takeoff crash of a torpedo plane. The plane roared down the deck in normal fashion, flew steadily over our bow, started to climb, then unaccountably dipped its starboard wing almost vertically and spun into the sea, only a few hundred feet from the ship. Breathlessly we waited to see whether the three men aboard could get clear before it sank. None did. Our plane-guard destroyer rushed to the scene, but even as she arrived, a muffled explosion signaled the detonation of the airplane's depth charges, and all that remained was a widening patch of debris and oil.

The next night the ship was to load ammunition for the following day's gunnery practice. In the darkness on the flight deck, a seaman stepped backward to steady himself and fell to his death down the open well that housed the ammunition hoist. In the morning, we heard for the first time over loudspeakers in every part of the ship the most solemn of all the boatswain's calls, "Now, all hands, bury the dead."

On the heels of our tragedies came our disappointment. We had long looked forward to completing our altitude runs to calibrate our radar, but we found that it was simply impossible to do so in the narrow, land-locked gulf. Radar echoes from the mountains were so strong and extended over such a broad segment of the horizon that we could not track our aircraft far enough to do the job. Once again we had to plead the importance of the project, and it was to the credit of the air and executive officers that they listened patiently and allotted us time on the return passage. Thanks to the eventual completion of the job under scientifically controlled conditions, we were generally able to estimate altitudes in the air battles to come far more accurately than some other ships in our task force. Even when altitude-determining radar reached the fleet a year and a half later, our estimates were almost as close as those of the ships that were fortunate enough to receive it. On the way back from Trinidad, we felt much more like a fighting ship than we did on our way down.

The relationship between the regulars and the reserves was never a problem. The reserves were eager to learn, and the regulars were glad to share their knowledge and to teach. As the pilots came to know the ship's officers and vice versa, through daily contact in the wardroom as well as in the areas where their duties brought them into close contact, friendships were formed and a mutual respect began to develop. The same was true of the interplay between the various division officers on the ship: V-1, the flight deck; V-2, the hangar deck; V-3, radar plot, air plot, aerology, air combat intelligence and the photo laboratory; gunnery, navigation, engineering, etc. We came to understand how none of us could act without the others, and to appreciate the teamwork that an aircraft carrier epitomizes. Working across all these organizational lines, the medical staff and the chaplain were sources of invaluable help to all in time of need.

Off Norfolk, the air station sent a drone out to sea to provide us with target practice. These first models of pilotless aircraft were slow and clumsy by modern standards, but they nevertheless presented a challenge. We picked ours up on the radar when it was many miles away, passed its range and bearing to the gunnery officer, and when it came within range, the gun crews knocked it down on its first pass. The captain was pleased, and took the occasion to release all restricted men on board, so that they too could enjoy liberty when we reached Philadelphia for our finishing touches before leaving home in earnest.

As we came to our old berth in the Philadelphia Navy Yard on a cold November afternoon, I think we all felt for the first time that we really had become a part of what the Navy likes to call "a force in being." Fred Jones and I went over the side together, pleased at the progress of our ship and her air group, and excited at the prospect of five days' leave in Columbus, Ohio, where Peggy and Nick went when the ship left Philadelphia. Like Dutch, both Fred and I believed that the next cruise would be "the long one."

Upon our return to the ship in early November, her tempo had noticeably quickened. Yard workmen were still very much in evidence, making final adjustments to gear that had not performed quite satisfactorily on the shakedown cruise, and installing final items of equipment before our departure to join the fleet. Where their pace had been leisurely six weeks before, it now was hurried, as if they sensed that time was running out and that the job must be finished now or never. The crew also had stepped up its pace. Jobs that had taken all day before were now being done in a few hours.

A pier-mate, the French cruiser *Georges Leygues*, had joined us. As one of the French warships that had eluded German capture or internment, she had been placed under Allied command and had come to Philadelphia for refit and modernization. She was separated from us only by the width of the pier, and considerable interchange, mostly friendly, took place between her crew and ours. Once, during a watch as officer of the deck, I noticed that the biggest wooden hogsheads I had ever seen were being laboriously rolled up to the foot of her brow, and that her officer of the deck had left his post and was holding a long conversation with the French seamen standing in a group around the nearest hogshead. As soon as I was relieved, I went over to see what was going on. The casks were full of wine, and, with obvious relish, the OOD was sampling each one, "to see if it is fit for our receipt."

A British cruiser was moored nearby, and before long the new American battleship *Iowa* steamed in from one of her first trial runs. Together with various destroyers and small craft of all descriptions, we made a cosmopolitan and businesslike-looking group.

When the day came for our departure, there was no fanfare. We simply slipped quietly away from the pier, picked up the two destroyers that were to accompany us, and headed for sea.

VIII
Return to Pearl Harbor

I asked Tom Sorber, "what is your guess as to where we are going?"

"Truk," he replied, naming the most fearsome place he could call to mind.

There was no doubt in any of our minds that we were headed for the Pacific, but speculation ran rife as to whether we would be assigned as convoy escort like the small carriers, or to a striking group like the large ones of the *Essex* class.

All the way down the East Coast, we made a fast passage, with our combat air patrol of fighters and our antisubmarine patrol of torpedo planes aloft all during the daylight hours. In our screen up ahead, the two destroyers rolled and pitched in unison, and I was pleased to note that one of them was named the *Lewis B. Hancock*, in honor of the late husband of my old friend Joy Bright Hancock. It felt almost like a tie with home.

We arrived in Cristobal at the Atlantic end of the Panama Canal in the late afternoon of a December day, and dropped anchor for the night to await our transit of the locks the next morning. Liberty ashore that night took on some of the aspects of New Year's Eve in a madhouse. For months afterwards the crew relived and embellished their moments of triumph with "the Blue Moon girls," the hostesses of Canal Zone cafés with a worldwide reputation for the warmth of their hospitality to sailors, their ability to drink small glasses of colored water all night and make it seem like whisky, and their skill in corralling the dollars of every sailor who went ashore.

In the morning, with half our complement wishing that it were already dead, we began our transit of the canal in a tropical deluge. As the donkeys inched us through the first locks, word spread throughout the ship that the new *Essex*-class carrier *Intrepid* was a few miles ahead of us, and that we would probably rendezvous with her at the Pacific end of the canal. An hour or so later, in the middle reaches of the canal where the channel twists and turns in a series of tight curves, we came suddenly upon her, ingloriously fetched up on a mudbank.

"Her pilot must have been in our liberty party last night," said Dutch.

Whatever the cause, she was not able to proceed with us, and we heard later that she had to be laid up for several weeks of repairs before she could again start out to join the other carriers in the Pacific Fleet.

Once past Panama, we ran up the coast of Central America and put in at historic North Island in San Diego Harbor. Here, our orders were to load all the Marines and all the aircraft we could carry, and transport them to Hawaii. The aircraft came on board first, and it took us a whole day to hoist them up to the flight deck, lower them down the elevators, and secure them to the hangar deck. Finally, we packed the flight deck solid with row after row of fighters, dive-bombers and torpedo planes.

In the late afternoon, the executive officer passed the word to all the officers that anyone going ashore would not be permitted to leave North Island, and that the ship would be leaving early in the morning. During the night, our Marine passengers, one thousand strong, came aboard and spread their blankets wherever they could find room enough to stretch out: in the catwalks, under the wings of aircraft parked on the hangar deck, and in every conceivable nook and cranny of the ship.

It was an uncomfortable passage for all concerned. In order to feed the men, the galley had to be kept in operation all day long and the chow line formed on the mess deck in one continuous shift. Below decks, cots, blankets, and bedrolls were everywhere. On the decks, aircraft were parked so tightly together that there was hardly room to squeeze between them.

Under these conditions, Captain Dillon began a high-speed run to Hawaii. Christmas was only a few days away, and we knew that he would try to make Pearl Harbor in time to celebrate it there if he could; he had been the commanding officer of the Naval Air Station at Kaneohe Bay and had many friends throughout the islands. In three and a half days, we had the island of Oahu in sight, and before sundown on the day before Christmas, we steamed into the narrow entrance to Pearl Harbor and made fast to the pier in the navy yard. The Marines immediately went over the side to quarters ashore, and the ship's company once more settled down to normal living.

Christmas dinner in the wardroom, with turkey and all the trimmings, passed pleasantly enough. Then came orders for me to leave the ship immediately and join the USS *Yorktown* for a practice cruise during which fighter direction exercises were to be held. This did not present any novelty, after my cruise in the *Essex*, but it was interesting to see how they did things aboard Joseph J. (Jocko) Clark's "Fighting Lady." They certainly did them well.

Chuck Ridgway was her fighter director, and he had much practical advice to offer in connection with task-force operations involving large groups of carriers. The *Yorktown* had recently returned from supporting the invasion of Tarawa, and her approach to within a few miles of the beach was still plotted on the dead-reckoning tracer in her radar plot. It was sobering to learn that the carriers were operating right up to the approaches to the islands, but comforting to hear that they were concentrated in strong task groups with the support of battleships, cruisers, and many destroyers.

With much food for reflection, I returned to the *Langley* a few days later to find that Dutch had assigned me, on a permanent basis, to the job of intercept officer in radar plot. This meant that I would hold a key position on our team, and would be the one who actually vectored our fighters to intercept the enemy.

IX
The Capture of Kwajalein, Majuro and Eniwetok

Early in January 1944, momentous changes were made in the organization of the Pacific Fleet. Admiral Towers was relieved as ComAirPac and promoted to Deputy CinCPac, still not the seagoing command he yearned for, but clearly Admiral Nimitz' right hand. The *Essex* and *Independence*-class carriers were organized into Task Force 58 for the first time, along with fast battleships, escorting cruisers and destroyers. Admiral Raymond A. Spruance was put in command of the Fifth Fleet and Admiral Marc A. Mitscher was assigned command of Task Force 58, the fast carrier task force. A scheme of command rotation was worked out: while Spruance and Mitscher were at sea carrying out one series of operations, the team of Halsey and John S. McCain was planning a subsequent series and preparing to take over for its execution, at which time the same ships would be designated the Third Fleet and Task Force 38.

For Operation Flintlock, the assault on Kwajalein, Task Force 58 was organized into four task groups, each with three carriers. Groups normally steamed as one unit, but were positioned just far enough away from one another to allow sufficient air space for the launching and landing of their aircraft in accordance with their own schedules. In total, the task force could launch 650 aircraft; and, with its 8 battleships, 6 cruisers, and 35 destroyers escorting the 12 carriers, it was an awe-inspiring sight.

Each task group was commanded by a rear admiral. The *Langley*, along with the famous old carrier *Saratoga* and our sister ship, the *Princeton*, was assigned to TG 58.4, under Rear Admiral Samuel Ginder. The fighter directors of the carriers were summoned to the *Saratoga* for a preoperation conference, and, as I went up her gangway, I could feel the impact of all the history that had been made aboard her. One of the points emphasized at the conference was that, for the first time, we would be attacking a parcel of territory that was Japanese before war started. Consequently, difficult as the taking of Tarawa had been some three months previously, it was presumed that Kwajalein would be even better fortified and even more stoutly defended. Whether or not the Japanese fleet chose to come out and fight, it was to be expected that very large numbers of their land-based aircraft would be staged down through the islands to attack us.

On 19 January 1944, the great task force sortied from Pearl Harbor and headed southwest toward its target, more than 1,500 miles away. Kwajalein, some sixty miles long by thirty miles wide, is the largest atoll in the world. A short flight to the east is its neighbor, Wotje, to the southeast Majuro, and somewhat farther to the northwest lies the atoll of Eniwetok. For Flintlock each of the four task groups was given a specific part to play, and ours in TG 58.4 was to attack the airfields of Wotje, Taroa, and Eniwetok and deny their use to Japanese reinforcements.

After approaching the enemy atolls under strict radio silence and hoping for surprise, we launched our first strikes at dawn on 29 January. Wotje was the first target for the *Langley*'s aircraft, and Ed Konrad, our air group commander, led our strike group to join up with Jumping Joe Clifton, the strike leader from the *Saratoga*. We were some seventy-five miles offshore and could track them all the way until they dove to attack the runways and hangar facilities on Wotje. We thought we could detect a few enemy aircraft over the target, but none came out to challenge us, and the returning strike reported only a very few in the air or on the ground. Antiaircraft fire, however, was intense.

All day long, the *Langley* sent in the torpedo planes and fighters of Air Group 32, loaded with bombs to pockmark the runways and destroy the hangars, oil-storage facilities, and gun emplacements. Meanwhile other task groups were doing the same on Kwajalein and scouting the landing areas on Majuro, which put up almost no resistance despite the fact that it offered the best anchorage in the Marshall Islands.

At night, we pulled farther away from shore and braced ourselves for an air attack that did not materialize. At dawn next day, we repeated the performance and kept up the pressure with strike after strike throughout the day. Then, on the thirty-first, the Marines and Army troops landed at Kwajalein and Majuro to meet bitter fighting at the former and only token opposition at the latter. Still no aircraft came out to attack us, and after continuing to provide air support for the next three days, we shifted over to Eniwetok and made our first strikes on the airfield there.

After a week of bitter fighting on Kwajalein, the atoll was declared secured, and shortly afterward we steamed into its lagoon aided by a roughly drawn chart prepared on the spot by a minesweeper exploring the passages between the islets linked by the reef. As we went in, we passed close by still-smoldering, wrecked tanks on the beaches. The devastation was frightful, the main island virtually denuded by what came to be known as "the Spruance haircut." As I stood on the flight deck after we had made our entrance, a radarman standing next to me looked at the level sandspits all around us and exclaimed, "All that, just for this?" I think he spoke for us all.

After only about eight hours in the Kwajalein lagoon, we were ordered to proceed immediately to Eniwetok. In view of the lack of air opposition so far, the decision was made to speed up the timetable and capture Eniwetok right away rather than several weeks later, as had been planned. It was judged that one task group could handle the softening-up of Eniwetok, and that assignment fell to us in TG 58.4, releasing the other three groups to put to sea for the first carrier raids on the fortress of Truk.

By 10 February we were back on station at the oval atoll of Eniwetok, striking both the main islands, Engebi to the north and Eniwetok to the south.

No enemy aircraft appeared to interfere with our bombardment. For this reason and because antiaircraft fire had ceased and there were no signs of activity, some of our pilots were convinced that the enemy had evacuated the atoll. This was far from the case, as the Marines and the Army discovered when they landed on Engebi on the eighteenth and on Eniwetok the nineteenth. Apparently the Japanese had decided to husband their ammunition and to avoid revealing where their guns were sited.

It was part of our mission to provide air support for the troops once they had landed, and we spent the next week launching strikes against tactical targets identified by the ground commanders. For the first time, we were able to monitor the radio frequencies used by the tanks on shore and to follow the course of the battle on grid charts. A chart of the lagoon had been captured at Kwajalein, which was fortunate for everyone because, here again, the US Navy had none of its own, and in prewar years the Japanese had been extremely secretive about all the islands mandated to them. With Marines landing on Parry Island in the atoll on the twenty-second, the end was in sight, despite stiff resistance both there and on Eniwetok.

We remained on station to protect against raids from other island bases until forces were well established at the airfield on shore, then headed for Majuro, relieved that the assaults had been successful, and very puzzled by the lack of air opposition.

While we were busy at Eniwetok, supply ships and tankers streamed into the beautiful anchorage in the Majuro lagoon, ready to rearm, refuel, and reprovision us. The operation provided us with an example of the miracle of logistics conjured up by the Pacific Fleet. We had been at sea for two full months since our departure from Pearl, and now for the first time there were packets of mail awaiting us. In my bundle there were letters from Uncle John, from Peggy, and from Rud Platt, all advising me of my mother's death in California some five weeks earlier. Sadly, I thought of how cheerful she sounded when I called to say goodbye from North Island in San Diego, and how much I wished I could have done more for her in her illness. But I was thankful that she would suffer no more and for the love and care that Uncle John had given her in her last years. As best I could, I answered the letters and just as I had finished the last reply, the chaplain appeared at my door.

"John, I have a sad message for you from the Red Cross."

"I know, Padre, I've already had the word."

"If you want me to, I'll gladly recommend you for emergency leave."

"No thanks. I know my mother would want me to get on with the war."

For days afterward I had a lump in my throat, but was convinced that my decision to stay with my ship was the right one. There was simply nothing I could have done at home that had not been done long since. There was nothing for it but to get on with the next operation.

X
Espiritu Santo and the First Strikes on Palau

The next few days were busy ones. Our bomb-stowage lockers were virtually empty as a result of all the strikes we had launched against Wotje, Taroa, and Eniwetok; we needed to replace damaged aircraft; and we had to reprovision and refuel. Within a week, however, our task group weighed anchor and departed for Espiritu Santo in the New Hebrides, some 1,500 miles south.

On our first approach to Kwajelein in January we had crossed the international dateline, and thereby qualified for the Navy's Golden Dragon fraternity. Now, we would cross the equator, and those of us who had not done so before would become "shellbacks." Captain Dillon authorized the crew to rig the hangar deck for the ceremony. Under the direction of a chief boatswain's mate, a greased slide into a water tank, complete with paddle stations en route, was prepared, and officers and enlisted men alike were subjected to various forms of hazing. After two months of concentrated business, all hands enjoyed the fun, which served the purpose of blowing-off steam.

We suspected that we were being sent far enough south to be part of an assault on Rabaul. General Douglas MacArthur, however, had sent a strong reconnaissance force into the Admiralty Islands, off the northern coast of New Guinea, and found that they were very lightly defended. In taking them, he acquired a fine fleet anchorage at Manus, and even more important, a platform from which both Rabaul and Kavieng could be neutralized, thereby avoiding the heavy casualties that their capture would have entailed.

With this development, the high command authorized another advance in the Pacific timetable, namely the leapfrogging of eastern New Guinea to seize Hollandia, far to the west. One of the factors menacing such a long leap forward was the major Japanese air and naval base in the Palau Islands, just east of Mindanao. After the severe raid that the other fast carriers launched against Truk while we were at Eniwetok, many of the Japanese units moved west to Palau, where they had a fleet anchorage and two airfields. It was decided that these would have to be put out of action before troops could be put ashore at Hollandia. The *Saratoga* was detached from TG 58.4 and sent to the Indian Ocean. Task Force 58 was reorganized into three groups and we were placed in TG 58.3 under Admiral Ginder along with the *Yorktown*, *Lexington*, and *Princeton*.

On our long passage to Espiritu we were in little danger of air attack, and, except for routine combat air and antisubmarine patrols, we had little to do. One day when I was off watch, I was talking to Tom Smith, the air plot officer, in the wardroom and found that he knew Espiritu Santo well. Tom was one of three advertising men in the ship. He had worked at Young & Rubicam in New York in a job similar to mine and, along with McKee Thompson from McCann-Erickson, we had fun in the evenings discussing the business and composing mythical commercials to be run over the ship's loudspeaker system. Now it developed that Tom was in the *Hornet* when she was sunk by enemy aircraft in the vicinity of Guadalcanal. He was providentially rescued by a destroyer, herself under air attack, and landed on Espiritu. The base commander there had seen to it that he was supplied with new uniforms and assigned to a tent for what was presumed to be a short wait for orders from CinCPac. When two weeks had passed and none had arrived, Tom asked for a temporary job and was assigned to serve as a harbor pilot. After six months, certain that he was one of the forgotten men of the Pacific, he wrote a personal letter to the former executive officer of the *Hornet*, Apollo Soucek, whom he learned had also been rescued and was then at ComAirPac. When Apollo, who was Peggy's cousin, got the letter, he immediately sent a dispatch to Espiritu, ordering Tom home for long-overdue survivor's leave, and subsequently to the *Langley*. Small world!

Espiritu appeared to be everything a South Pacific island should be: lush, green, densely tropical, and with a languid climate. For the first time since we left Oahu in January, I stepped ashore to attend a party at the Crow's Nest Officers Club at Teteron Bay. During the festivities, which went on for quite a while, a group of us from the *Langley* negotiated the lease of a 60-foot, high-speed Army crash boat from a young Army lieutenant. The duration of the lease was from 0800 the next morning until 2000 that night, and the price agreed upon was two quarts of bourbon. Luckily, the "wine mess" that we had established in Philadelphia allowed us to unlock the liquor compartment and take some ashore when we were in port.

By arranging substitutions on the next day's watch schedule, about ten of us were able to take advantage of this opportunity, and took the boat over to Aobe Island, perhaps forty miles away. If Espiritu was the inspiration for James Michener's *Tales of the South Pacific*, as many later believed, then surely Aobe was the model for Bali Hai. With beautiful beaches, tall palms, verdant growth, and misty mountains in the background, it was a dreamlike oasis in the world of war. The natives lived in well-built thatched houses, shaded by the palms and cooled by the sea breeze. Anything they planted seemed to grow: bananas, papayas, mangoes, pineapples, and varieties of fruit we had not seen before. The beaches were strewn with shells including cat's-eyes as big as silver dollars. The natives fished from the shore with casting nets and offshore from sturdy canoes hollowed out from coconut logs and fitted with outriggers.

We spent a delightful day exploring the island on foot and chatting with the natives as best we could, largely in pidgin. We gathered that they knew there was a war going on, but seemed wholly ignorant as to why or who was involved. We wanted to load up with fresh fruit of all kinds, and they were eager to trade for undershirts, which apparently were at a premium as shields against the sun. When it came time to leave, about fifty natives, including well-fed, healthy-looking children, came down to the beach to see us off. For a long time afterward we thought back on that day as the nicest day of the war.

On 23 March we set out to rejoin the other task groups at a refueling rendezvous along the way to Palau. Once we passed Guadalcanal and Bougainville and headed west, past New Ireland and the Admiralties, it was like running the gauntlet because we were between the northern coast of New Guinea and the seemingly endless string of the Carolines, all in Japanese hands. The east-west axis of this corridor extends for nearly 1,500 miles, all within range of Japanese search planes from airfields in the Carolines and New Guinea. It was inevitable that the enemy would spot us, and a search plane from Truk did so early in the game.

On 26 March we rejoined the other task groups and refueled while underway at a rendezvous with tankers east of Kavieng. We then turned west and plowed ahead despite the frequent presence of shadowing Japanese planes. One of these was first detected by our radar in the *Langley*, and we were immediately ordered to send out the combat air patrol to intercept it. The interception went smoothly and the fighters shot the snooper down, but not before it had come within thirty-five miles of the task force, which made us afraid that it had had time to get off another sighting report. Nevertheless, there was celebration in radar plot; at long last we were doing what we had trained so long and come so far to do.

After dark on 28 March, while we were still two days' steaming from Palau, the enemy attacked in force with torpedo bombers. In preparation for the attack and to silhouette our ships for the bombers, they dropped magnesium flares attached to parachutes, laid down float lights, and lit up the scene with pyrotechnics. Used as we were to cruising in tight formation with every ship completely blacked out, it was a shock to be suddenly exposed to the light of high noon and lined up like ducks in a shooting gallery.

At this stage of the war we were not operating night fighters in the task force, and our only defense against night air attack was our ship's guns. As the torpedo planes began their runs from the dark side of our force, to take full advantage of our silhouettes, we tracked them in mile by mile, sending ranges and bearings every minute or so to the bridge and the gunnery officer. When they approached our outer screen of destroyers, we saw the flash and heard the ear-splitting crack of the 5-inch 38-calibre antiaircraft guns carried by all ships except the CVLs. As the planes came closer and closer, the cruisers and battleships inside the screen and the *Essex*-class carriers opened up. When these proximity-fused shells came close enough to an attacking plane to explode, the plane itself would turn into a torch and plunge in a parabola into the sea. Any plane that survived the 5-inch barrage and pressed on was then subjected to a hailstorm of fire from every ship's 40-millimeter, 20-millimeter, and 50-calibre fire. This was the point at which our own ship's gunners came into action and the noise was overwhelming.

Our own pilots, off duty for the night, reacted to the Japanese raid like college boys at a football game. A large group of them went to the forecastle where they could watch the attacks from under the projection of the flight deck and criticize the tactics of the Japanese pilots. Interspersed with the gunfire, there were cheers and boos from the forecastle, depending on reaction to an individual effort on the part of our gunners or of the enemy's torpedo runs. Said Tom Sorber, "Next time, maybe we ought to sell seats to these guys!"

No ship was hit during the two hours the attack lasted, and a number of the attackers were shot down. By the morning of the thirtieth, we were close enough to launch our own strikes on Palau.

With all the warning they had had of our approach, the major Japanese warships at Palau had escaped by the time we arrived, but there were other ships in the harbor. These were promptly bottled up by mines dropped from planes from one of the other task groups, and when our attacks were over it was estimated that 36 ships, amounting to 130,000 tons, had been sunk or badly damaged.*

For three full days we launched continuous strikes at Palau, which was defended by its own aircraft based at Peleliu and by others from Yap. A great many were shot down by the fighters escorting our strike planes, and some by our combat air patrols. One Japanese bomber was "splashed" by our own ship's guns, and that afternoon our skipper received a blinker message from the task group commander: "Congratulations on losing your virginity!"

On its way back to the base at Majuro, TF 58 detoured to Woleai due south of Guam, but found few worthwhile targets there. On 6 April we were back at anchor in the Majuro lagoon.

By this time the various components of the *Langley*'s crew were beginning to have confidence in themselves and in each other. Our air group, AG 32, was hot, the ship had performed well in three operations, and we had come through our first serious air attack unharmed. It was a good feeling, and by way of celebration, the wardroom officers invited the captain to come to dinner with them. To lend a festive air to the occasion, we called on two great assets that came to us with Air Group 32. Lieutenant Commander Eddie Outlaw, skipper of the group's fighter squadron and later of the whole group, had organized a squadron jazz band and a vocal quartet known as "Outlaw's Angels." Both were surprisingly good, considering that they consisted entirely of pilots who had little time to practice. Above all, they were loud.

We were all very fond of "Gotch" Dillon. Short and a little bowlegged, he paced the deck almost like a bantam rooster; but he had a twinkle in his bright blue eyes, a great sense of humor, a pleasant Southern drawl, and an innate fairness in dealing with the men. Under Gotch, the *Langley* was a happy ship.

---

*Samuel Eliot Morison, *History of United States Naval Operations in World War II*, v. VIII (Boston: Atlantic Little, Brown, 1975), p. 32.

As Gotch entered the wardroom, we all rose to greet him and the band struck up, "He Don't Get Around Much Any More," a selection deemed appropriate considering that we were some five thousand miles from anything that could remotely be considered home. This was followed by the "Five O'clock Jump," "The Jersey Bounce," and other wild numbers of the day; and after dinner the Angels favored us with their spirited theme song, "Who Shagged O'Reilly's Daughter?" It was a tossup as to who had the best time, we or Gotch. It is a sad fact of Navy life that the captain dines in lonely state in his own quarters. He dines in the wardroom only on special occasions, and then only on invitation from his officers.

We spent the next week rearming and reprovisioning, jobs that were demanding and difficult for the crew, whether they were accomplished when the ship was at anchor or, as they were a little later on, when she was underway. In either case, bombs and torpedoes had to be gingerly hoisted aboard and stowed extremely carefully; and case after heavy case of foodstuffs and other stores had to be manhandled to their respective storage areas. In some ways it was the hardest work of all for the crew.

XI
The Assault on Hollandia

For Operation Reckless, the seizure of Hollandia on the northern coast of New Guinea, more changes were made in the fast carrier task force. The high command was critical of Admiral Ginder and decided to transfer him from combat operations to command of a group of escort carriers in the Service Force. He was relieved as our commander in Task Group 58.3 by a veteran rear admiral, J.M. (Blackjack) Reeves. The group's carrier strength remained at four: the *Enterprise*, the *Lexington*, the *Princeton*, and the *Langley*. Another change saw Jocko Clark, the erstwhile skipper of the *Yorktown*, fleeted up to rear admiral and placed in command of TG 58.1.

Hollandia was basically General MacArthur's operation. While we were taking out the Japanese air forces at Palau, he had sent General George C. Kenney's fighters to the airfields around Hollandia, and it was thought that the two operations had left little in the area with which the Japanese could oppose MacArthur's landing forces. Our assignment in Task Force 58 was to patrol offshore and intercept any aircraft the enemy might stage down through the Carolines, and fly softening-up strikes before the landings on New Guinea.*

---

*My feelings toward General MacArthur and the Army troops under his command contrasted sharply with my feelings toward the Marines. While I shared the widespread Navy view that MacArthur was a show-off and a blowhard, I could see that he did have some propaganda value for our side, and I was willing to concede that he got his troops where they were supposed to be. In the most difficult situations, however, I preferred every time to be working in support of the Marines for whom I developed an admiration akin to awe. In operations where both Marines and Army troops were similarly engaged, it seemed to me that the Marines invariably emerged as the more effective force.

We left Majuro on 13 April, made a brief feint in the direction of Palau, and struck at Hollandia on the twenty-first. Our ships steamed close to the shore, the mountains being clearly visible, but, other than the occasional "snooper," or search plane, very few runs were made at us. Reinforcements must have been flown to Hollandia after General Kenney's raids, because our pilots found many aircraft on and around the New Guinea fields. Strangely, few of them took to the air, and most were destroyed on the ground by 50-calibre strafing.

Night fighters were launched from the task force for the first time, and there was work for them to do. Among their missions was flying over Japanese troop concentrations to keep them awake at night. One night after our troops had landed, we directed a fighter onto the tail of a formation of enemy bombers flying from west to east along the coast, toward the site of our landings. When this fighter got close enough to "lock on" with his own radar, he did a superb job of shooting down first one, then a second, and finally a third before the formation could get to its point of bomb-release.

Next day two of our fighters returning from a strafing run spotted a downed pilot in a rubber raft. They called us to say that they would orbit over the raft, which was about eight miles off the coast, so that we could get a good radar fix on its position. With this information in hand, one of the cruisers in our group launched a float plane to fly the rescue mission. Our fighters remained in orbit while the float plane was en route, and Dick May, the leader of the section, spotted a Zero coming down to strafe the raft. He and his wingman powerdove onto the Zero's tail and shot it down just as it opened up with its first burst at the raft. Soon afterward the cruiser's plane landed on the water, taxied over to the raft, and recovered the pilot unhurt.

We never did discover why so many aircraft on the Japanese fields failed to take off. Speculation ran high that it was because all their gasoline had been destroyed. On the twenty-fourth, we went to Manus for refueling and new orders.

In order to deny the Japanese the use of Truk as a base from which to attack our new foothold at Hollandia, our next assignment was to strike it with all twelve carriers. Even though the other three task groups had dealt it severe damage while we were assaulting Eniwetok, we were extremely wary, for there had been plenty of time to rebuild its strength, and it was, after all, the "bastion of the Pacific" for all Japan.

Under the strictest radio silence, we steamed north from Manus, in the Admiralties. The attack plan called for the *Langley* to launch a fighter sweep of eight Grumman Hellcats from VF-32 to clear the sky over Truk of enemy fighters before our bombers and torpedo planes arrived on the scene. Accordingly, before dawn on 29 April, Eddie Outlaw led the eight planes in. While he was en route, we could see on our radar that there were many bogeys circling over the various islands in the Truk complex, and were able to warn him that he would meet plenty of opposition. When he sighted them, he found them to be a large group of Zeros, some sixty in all. He ordered his two divisions to turn on the newly developed water-injection system that Pratt & Whitney had invented to give an engine a momentary burst of extra power for use in emergencies. (This was the first time our pilots used the system in combat.) His eight fighters then dove to the attack, and in approximately fifteen minutes of furious combat, shot down twenty-one Zeros and damaged several others.* Outlaw himself shot down five; his wingmen, Dagwood Reeves, Hollis Hills and Dick May each got three.

Later in the morning, when they were sure of the carriers' location, the Japanese launched a bombing and torpedo-plane attack on us, which we were able to intercept with the combat air patrol and then drive off any survivors with antiaircraft fire. Meanwhile, our own attack forces were strafing enemy aircraft still on the runways and bombing every ship and significant shore installation they could find. Unlike most of the atolls in the Pacific, Truk's islands rise out of the lagoon itself and are surrounded by the barrier reef instead of being linked to the reef like beads in a necklace. This configuration allowed the Japanese to install antiaircraft guns on the reef itself, as well as on all the main islands, such as Dublon, Moen, and Tol. Their fire was murderous. In exchange for two days of attacks on Truk, resulting in the crippling of the base for the rest of the war, Task Force 58 lost twenty-six planes along with forty-six airmen.

More than half of the downed airmen were rescued by the heroic efforts of the submarine *Tang* working in tandem with a Chance Vought OS2U Kingfisher float plane launched from one of the cruisers in our task force. The *Tang* was the submarine assigned to patrol between the task force and the target area expressly to rescue pilots who made water landings. In this operation, it rescued no fewer than twenty-two aviators and crewmen. At one point it fell to the *Langley*'s fighters to provide air cover for the *Tang* as it came in very close to the

---

*Langley Action Report of Air Assault on Truk, 29-30 April 1944.

Top: Downed pilots being rescued by the Vought Kingfisher float plane.
Bottom: The survivors are transferred to the submarine *Tang*.

reef to pick up one of our own fighter pilots, Lieutenant Barber, who was in a rubber raft. Because, understandably, the surfaced submarine was reluctant to come within range of the guns on the reef, Eddie Outlaw organized a continuous strafing of that section of the reef. Then I heard him call the submarine and say in his best North Carolina drawl, "Come on in here, buddy, and pick up this boy. Ain't nobody gonna hurt you!" Meanwhile, the Kingfisher landed on the water and taxied from raft to raft picking up as many other pilots as he could. Being a low-powered, two-place plane, all the Kingfisher could do was to provide the pilots with something to hang onto while it laboriously taxied between the reef and the surfaced submarine. On one of its journeys, six aviators were perched on its wing. Barber made the long voyage back to Pearl Harbor in the *Tang*, and returned to us by air to rejoin his squadron.

With no substantive targets left to attack, we retired to the anchorage in Majuro with a great sigh of relief.

XII
Operation Forager and the Battle of the Philippine Sea

The main islands of the Marianas group are Saipan and Tinian at the northern end, Rota in the middle, and the larger island of Guam at the southern end. They lie approximately midway between the Carolines to the south and the Bonins to the north. Their capture became one of the major objectives of the Pacific war, both to protect our flank during our westward advance to the Philippines and, equally important, to serve as a base from which the new B-29 bombers of the Army Air Forces could reach Japan proper. A further incentive was the fact that Saipan was the command center for all Japanese forces in the central Pacific. The American planners of Operation Forager considered it certain that the Japanese would at last order out their battle fleet to attack us and defend the Marianas.

In the six months since we had left the United States the shipyards at home had gone into high gear, and the Navy had sent out powerful additions and replacements to Task Force 58 in time for the Marianas invasion and whatever subsequent reaction the Japanese cared to mount.

When the fast carrier task force sortied from Majuro on 6 June 1944, it presented an incredible sight: 15 aircraft carriers bearing 788 planes, 7 modern high-speed battleships, 10 cruisers, and 60 destroyers. Manning this armada were 98,618 men,* and behind them, in other ships, would come 127,500 troops from as far away as Guadalcanal, Eniwetok, and Pearl Harbor, the last 3,500 miles distant.**

---

*Morison, p. 258.
**William T. Y'Blood, *Red Sun Setting* (Annapolis, Md.: Naval Institute Press, 1981), p. 30.

Once again we had a change of admiral and were shifted to Task Group 58.4. Along with the *Essex* and the *Cowpens*, we were to be under Rear Admiral W.K. Harrill. There were many changes in personnel, and I was glad to see that Ralph Ofstie, a friend from my CinCPac days, now had command of the *Essex*. In the fighter director chain of command, I recognized Lieutenant Joseph R. Eggert, the task force FDO; Lieutenant C.D. Ridgway, FDO in the *Yorktown* and now Task Group 58.1 FDO; and our own task group FDO, Lieutenant Commander F.L. Winston, with whom I had served during the training cruise of the *Essex*.

The plan called for a huge fighter sweep to be made simultaneously over all the main islands early in the afternoon of 11 June. Our group's target was Saipan, and the *Langley*'s fighters led by Outlaw again, proceeded to strafe and bomb the many targets around the airfields and the towns of Garapan and Charen-Kanoia. In the course of these attacks, dense clouds of black smoke began to form, augmented by what apparently were smoke pots intended to obscure the pilots' vision. As they dove down through the smoke, two *Langley* pilots were shot down by the vicious antiaircraft fire at all key points: one of them was Outlaw's wingman, Lieutenant (j.g.) Donald E. Reeves. A fine pilot and a great shipmate, Dagwood was never seen again. Our other casualty was Lieutenant M.M. Wickendoll, who fortunately was able to make a water landing and was picked up by the destroyer *Thatcher*.

With the Marine landings on Saipan scheduled for 15 June our mission continued to be to sweep enemy aircraft from the air over the Marianas and to seek out and destroy as many enemy gun positions as possible. In the course of strafing the island of Pagan on the twelfth, our task group's strikes discovered a Japanese convoy approaching the area with reinforcements for the Marianas garrison. For two days the dispatch of this convoy took on top priority, and by the end of the thirteenth we had sunk ten transports and four of its escort vessels. Meanwhile all the principal islands in the chain were receiving a heavy going-over by the fighters and bombers of other task groups, and enemy air opposition was soon dissipated.

Once the Marines had fought their way ashore, our aircraft provided close air support during the days of furious resistance. Not until the seventeenth was that job turned over to the escort carriers that accompanied our transports.

Admiral Mitscher, commanding the fast carrier task force, decided to send two of his four task groups northward to attack the island of Iwo Jima in the Bonins. The purpose of this was to keep the Japanese from using it as a staging point for airplanes en route to the Marianas from the home islands of Japan. The two groups chosen were ours under Admiral Harrill and 58.1 under Jocko Clark. We in the *Langley* did not know it at the time, but apparently Admiral Harrill had very little appetite for this dangerous sortie so close to Japan, and attempted to beg off on the grounds of being low on fuel.* Mitscher did not accept Harrill's argument and, along with Clark's group, we turned north in ever-worsening weather.

On the evening of 14 June, Admiral Spruance informed both Harrill and Clark that at long last the Japanese fleet was coming out to fight. Consequently, the two task groups were to spend only one day striking Iwo, Chichi, and Haha Jima, and then rejoin the task force immediately in preparation for the approaching battle.

In high winds and heavy seas, we launched our first strike on Iwo Jima under heavy cloud cover. As we were making our approach we spotted a Japanese sampan on line with the target and promptly sank it, but probably not before it could get off a warning message. Our first fighter sweep over the island found thirty-seven Zeros airborne and waiting, and twice as many more based on the airfield. In addition to the heavy air opposition, our fighters and bombers alike ran into the heaviest antiaircraft fire yet encountered, and when they returned to the carrier, she was rolling and pitching so heavily that there were many wave-offs, missed arresting wires, and barrier crashes. Hal Paine, the air combat intelligence officer attached to our VT-32, had been given permission to fly as an observer in the tail gunner's seat of a TBM. When his plane landed back on the *Langley*, Hal was dead. He had been hit by a shell burst. Pat Patterson, piloting a TBM, returned safely and made a good landing, despite having had his windshield shattered by shellfire and his face severely cut by dozens of glass splinters. After destroying many enemy aircraft we returned to the task force, as ordered, on the morning of 18 June. When we returned to Iwo much later in the war, we had a very healthy respect for its defenses.

The Japanese mobile fleet, counterpart to our fast carrier task force, and its supporting supply ships, was assembled at Tawi Tawi, just off the northeastern tip of Borneo. Under the command of Vice Admiral Jisaburo Ozawa, its main combatant ships were nine aircraft carriers,

---

*Y'Blood, p. 55, and Clark G. Reynolds, *The Fast Carriers* (New York: McGraw-Hill, 1968), p. 177.

five battleships, seven cruisers, and twenty-eight destroyers. These ships, which were organized into three task groups, represented the largest Japanese striking force of the entire war, but most of its pilots and aircrews were inexperienced and severely hampered by lack of training. On 13 June, Ozawa sortied from Tawi Tawi with orders to proceed to the Marianas and destroy the American invasion force. Almost from the moment of his departure, he was sighted and shadowed by US submarines, which provided the best intelligence Spruance was to receive for many days to come. After refueling at Guimaras in the Philippines, Ozawa passed eastward through San Bernardino Strait on the fifteenth, again sighted and reported by the submarine *Flying Fish*. It was now apparent that a major sea battle was imminent in the Philippine Sea, west of the Marianas.

Just as Spruance and Mitscher were poring over every report that reached the Fifth Fleet, Nimitz and Towers were doing the same thing at CinCPac. On the sixteenth, while we were on our mission at Iwo Jima, Towers warned Spruance and Mitscher that the Japanese might try to shuttle-bomb our carriers by launching strikes from beyond our round-trip range, then landing on their airfields in the Marianas to refuel and rearm for another attack on the long flight back to their own ships. This proved to be a remarkably prescient warning, for that was precisely the Japanese battle plan.

The decisions that Admiral Spruance made in the next two days reopened and reemphasized the old controversy between the "air admirals" and the "battleship admirals." Basically at issue was the contention of carrier commanders such as Towers, Halsey, and Mitscher that the fast carriers should not be tied to supporting operations on the shore: they should be free to capitalize on their great mobility and striking power by ranging far afield to seek out and destroy the enemy's main forces. After the Battle of Midway, two years earlier, Spruance was criticized by some of the air admirals for not having pursued the remnants of the Japanese fleet more aggressively and persistently after the initial actions had been successfully completed. Instead, he had elected to turn back and disengage, staying relatively close to the scene at Midway. The air admirals contended that this election had lost us a great opportunity to destroy the Japanese naval forces once and for all, thereby shortening the war by perhaps years. Now, in the Philippine Sea almost exactly two years later, Spruance's tactics were to be subjected to the same criticism.

Spruance's main concern was to protect the integrity of the large forces already ashore on Saipan and of those still to be landed on Guam and Tinian. He was haunted by the thought that, if he took TF 58 away from the Marianas to find and attack the Japanese far to the west, a second enemy force might make an "end run" around him and create havoc at Saipan in his absence. Accordingly, he moved the American transports and supply ships clustered off Saipan to a position well east of the island to keep them out of harm's way, and tethered TF 58 to a defensive position just west of the islands. Rightly expecting to be attacked in this position early on the nineteenth, he ordered Mitscher to clear all carrier decks for maximum use of fighter aircraft by sending their bombers and torpedo planes to orbit out of the way over Saipan. He then disposed the four carrier task groups twelve miles apart from one another and pulled out the battleships from each group to form a fifth group fifteen miles to the west, in case the enemy should bring its ships within engagement range of surface forces. With the stage so set, the task force waited for the attack.

Proceeding east from the Philippines, Ozawa launched his strikes at maximum range, a few minutes before ten o'clock on the morning of 19 June. Large groups of bogeys began to appear on the radar at ranges of more than one hundred miles to the west. The combat air patrol was augmented by massive numbers of fighters, and those from the *Essex* in our task group were the first to engage the enemy, fifty-five miles away. The enemy planes behaved peculiarly. After approaching to within about sixty miles of our ships, they went into a circle and orbited for several minutes, apparently while their strike leader gave them final orders. This maneuver gave our fighters time to intercept them far from the task force and to get into optimum attack position. The pattern was repeated all day long, and many of the Japanese pilots indulged in aimless aerobatics before being engaged. The inexperience and lack of training of the enemy's pilots were apparent early in the day, and the four massive raids they mounted during the long day were almost totally destroyed. Few attackers were able to penetrate to our ships, and the *Langley*'s antiaircraft guns were not even fired, despite the 473 aircraft Ozawa launched against the task force. More than 300 were shot down over the task force, and stragglers who attempted to land on Saipan, Rota, or Guam were destroyed by fighters over the airfields. In all, it was a triumphant day for American fighters and fighter direction. In the Navy, 19 June became known as the day of the Marianas Turkey Shoot. It was at first thought that the fast carrier task force, in defending itself against the four main raids from the Japanese carriers and those of land-based aircraft from the islands, had shot down more than 400 planes. The elimination of duplicate claims later reduced this number to something over 300. By any count, it was the greatest single shoot-down of the war in any theater, including the Battle of Britain; and it was achieved with a loss of only 29 American planes, including six lost operationally.

It was also a great day for American submarines. Ozawa's force had long escaped detection by search planes from our task force, but was finally spotted by a patrol plane from Manus. Soon thereafter, the submarine *Albacore* got into position to attack the carrier *Taiho*, which Ozawa was using as his flagship. Several hours after a torpedo from the *Albacore* hit her, Ozawa had to transfer his flag to another ship, and the *Taiho* went down. A few miles away another submarine, the *Cavalla*, attacked and sank the carrier *Shokaku*.

For us in the *Langley*, the day was not so satisfactory. While we in radar plot had been rewardingly busy all day tracking and reporting the various raids and working with the fighters from other carriers to intercept them, our own fighters were thoroughly frustrated by being kept in reserve during much of the battle. Orders to keep them circling overhead in reserve were repeatedly received from the flagship, and we had no choice but to comply. By the end of the day, Outlaw and his squadron mates were fit to be tied. They could see the enemy's contrails filling the sky, but all too often they were not allowed to pursue them. They also could hear us directing the fighters of other carriers, and could not understand why we weren't using our own. Until they landed back aboard, there was no way we could explain that we were following explicit orders, and even then it was puzzling, particularly since the *Essex*'s fighters from the same task group were among the very high scorers.

The *Langley*'s VF-32 were not the only frustrated aviators. The night before the battle, Mitscher had sent a message to Spruance proposing that the task force head due west during the night and launch strikes at the enemy fleet at first light on the nineteenth, when, he estimated, the two fleets would be only 150 miles apart. Spruance rejected the suggestion and instead turned the task force back even closer to Saipan. Next day, with the wind from the east, we had to turn away from the enemy into the wind every time we launched or landed aircraft. Instead of launching strikes at the enemy, we spent all day defending ourselves against his strikes.

Operation Forager and the Philippine Sea   91

As evening came, Spruance at last turned the fleet west in search of the Japanese, whose position he still did not know with certainty. As he had done earlier in the month, when ordered to attack Iwo Jima, our task group commander, Admiral Harrill, complained that some of his ships did not have enough fuel left, and Mitscher ordered him to remain behind, refuel, and cover Guam and Rota, while the other three groups hurried west. This was the end of Harrill as a task group commander and, once again, we were in for a change of command in a very short time.

Although Mitscher had had search planes continually in the air since dawn, the enemy fleet was not spotted until the latter part of the afternoon of the 19th. Despite the 275 miles separating the two forces, Mitscher ordered strikes to be launched at once and warned all hands to be prepared for night landings when the planes returned. Once over their targets, the attacking planes did not have enough time or fuel left to execute their runs in a coordinated and orderly way. Nevertheless, they succeeded in sinking the carrier *Hiyo* and two oilers, damaging three other carriers, and shooting down 65 enemy planes. Their long flight back to their own carriers in the dark ended in a nightmare of water landings by planes running out of gas, chaos in the landing circle as planes desperately low on gas fought for priority to be taken aboard, and many wave-offs and barrier crashes. To help as much as he could, Mitscher ordered the carriers to turn on their lights, an unprecedented action for our forces, and even to shine searchlights as beacons for the homing planes, which were allowed to land on any carrier they could find. The destroyers did a heroic job of rescuing downed pilots and aircrews, but the toll was heavy. In all, 100 planes were lost and, while 160 men were rescued, 16 pilots and 33 crewmen were lost.

After shooting down a number of aircraft over Guam while we were refueling on the twentieth, we rejoined the task force the next day and flew searches for downed pilots as well as longer-range searches for the remnants of the Japanese. Ozawa was well on the road to escape in the direction of Okinawa. As at Midway, the Japanese fleet was grievously damaged but not annihilated.

Once again, our task group was ordered to return to the attack on Guam, while the other three were sent to Eniwetok to rearm and reprovision. On 29 June, Admiral Harrill came down with appendicitis and was flown to the hospital at Pearl Harbor. Rear Admiral Gerald F. Bogan took his place. On 9 July, Saipan was at last secured and we were well started on the systematic preinvasion assaults on Guam and

Tinian. The Marines and Army troops landed on Guam on 21 July, and the Marines stormed Tinian on the twenty-fourth in one of the best-planned and executed operations of the war. For all of July and early August, we remained on station in support of the troops ashore until resistance ended on Tinian on 1 August and on Guam a few days later. Only once did I have a chance to go ashore: we needed a spare part, and I took a boat to Saipan to get it. I inspected the sugar mill at Charen Kanoia, which our pilots had bombed; saw the prison stockade; and watched the shelling of Tinian across the water. By mid-August we were back at Eniwetok to prepare for the next operation.

XIII
Under Halsey to the Palaus and Philippines

With the Marianas secured at last, Admiral Spruance handed over command of the fleet to Admiral Halsey; the Fifth Fleet was redesignated as the Third Fleet, and Task Force 58 became Task Force 38. The fast carriers, sixteen strong, were regrouped. This time we, along with the *Essex, Lexington* and *Princeton,* were assigned to Task Group 38.3 under Rear Admiral Frederick C. Sherman. "Fearless Freddy" Sherman was to remain our leader through thick and thin for a long time to come, and we welcomed the continuity and aggressive spirit he gave to our operations.

After considerable debate within the high command, it was decided that our next objective would be the capture of the Palau Islands, in the approaches to the Philippines. Labeled Operation Stalemate II, this assault on a strongly fortified and strongly garrisoned group of islands necessitated far-ranging attacks on the Philippines themselves to prevent enemy air from counterattacking our troops once they got a foothold ashore. D-day on Peleliu was set for 15 September, and we sortied from Eniwetok on 28 August to undertake the preinvasion softening-up.

For three days prior to the landings, we struck at the visible installations of the main islands, and attempted to knock out obstacles to the planned landings on Peleliu, Angaur, and Ngesebus. Our fighters soon took care of the air opposition, but had a hard time finding targets in the interior, the enemy having concealed them with considerable skill. The Japanese commander had decided to hold the bulk of his 13,000 troops back from the beaches in heavily fortified positions inland. When these forces ultimately went into action on terrain of their own choosing, the fighting became ferocious and the American casualties very high: 1,950 killed and 8,500 wounded before the objectives were fully secured. Considering the high cost, many would have agreed with Halsey's view that we might well have bypassed Peleliu entirely. Fortunately, the largest island in the Palaus, Babelthuap, was bypassed, and its very large garrison was isolated until the end of the war.

After the full weight of the task force had been thrown against the islands between the sixth and ninth of September, Halsey left Admiral Ralph Davison's Group 38.4 to continue the work, and took the rest of us in the other three groups west to strike the airfields on Mindanao.

We had been briefed to expect massive air opposition from the scores of airfields in the Philippines proper, and were very surprised to find only token opposition at Mindanao. Halsey then took the task force north into the central Philippines, the Visayas. In quick succession, we struck Negros, Leyte, Cebu, Bohol, and Panay, sinking several enemy ships and destroying some two hundred of his aircraft. By 13 September, Halsey was convinced that the timetable for the invasion of the Philippines could be speeded up, Mindanao bypassed, and landings made on Leyte forthwith. He sent his recommendation to Nimitz, who got it approved by the allied heads of state, who were meeting in Quebec.

After Panay, we went north to Luzon and prepared to make the first strikes on the fortifications around Manila. In view of the tragic history of MacArthur's and General Jonathan M. Wainwright's defense of Manila and Corregidor in 1941 and 1942, this moment was full of emotion and excitement. On the evening before our first strike at Manila, 21 September, we could hardly believe our eyes when we went down to the wardroom for dinner. The chief steward's mate in charge of the wardroom mess was a Filipino with many years of service under the act of Congress passed after the Spanish-American War that made Filipinos the only aliens allowed to serve in the US Navy. This petty officer was quiet, efficient, and good-natured, a credit to the Navy. Manila was his home, and he had been quietly preparing a feast to celebrate the first attacks by our forces in the cause of liberation of the city. By scrounging and saving up supplies here and there, he had amassed the ingredients for a real Philippine curry, complete with chicken, rice, raisins, shredded coconut, ground peanuts, a delicious sauce, and even copious quantities of chutney. All this was spread before us that evening, a time when we had been at sea west of Pearl Harbor for nine consecutive months and accustomed to powdered milk and dehydrated potatoes. It seemed a miracle, and we gave the chief a standing ovation when it was over.

In the pilots' briefings before the first strike on Manila, the air combat intelligence officers stressed the need for as much information as possible about the prison compound at Santo Tomas near the university. Thousands of American civilians were interned there when the Japanese took Manila, and our sources didn't know what had happened to them. Our pilots were ordered to find the place, take pictures, and report any signs of life they could observe. When they returned from the first strike, they told us that, in the prison compound, they saw crowds waving them on, and their photographs substantiated the fact that many had survived their long imprisonment. Years later, our neighbor in the Virgin Islands, Spike Heyward, told me he was part of that crowd in the compound, as did Stanley and Bess Lehman, our neighbors in Lakeville, Connecticut. One can only imagine the joy with which they saw US Navy fighters sweeping in low over the hills and waggling their wings right over the compound.

All through the Philippines there were bogeys to be intercepted every day, as well as lucrative shipping targets. Hollis Hills, one of our fighter pilots, was shot down on a strafing run and made a water landing right in the middle of Manila Bay. Incredibly, he was rescued by a surfacing submarine and eventually returned to us. Day by day, the score of downed enemy aircraft mounted and, when we retired from the area at the end of September, a dispatch from Halsey noted that this was the first month in which the task force had destroyed more than one thousand enemy aircraft. He concluded in typical fashion with the observation "Not bad for a gang of old poops, young squirts, and lieutenant commanders!"

Our task group anchored for a day or two at Kossol Roads in the Palaus, then rejoined the other groups at the atoll of Ulithi, a magnificent anchorage taken almost without opposition on 23 September. We arrived on the fringe of a severe typhoon on 1 October and dropped anchor in carrier row for the first time at what was to become the major operating base of the Third and Fifth Fleets for months to come.

At Ulithi, in addition to refueling, rearming, and reprovisioning, we underwent two momentous changes: the relief of our greatly respected captain, "Gotch" Dillon, by Captain John F. Wegforth, and the relief of Air Group 32 by Commander M.T. Wordell's Air Group 44. The ship buzzed with activity. Dutch and I said goodbye to our many friends in AG 32 with whom we had worked for over a year. In the course of our farewells, John Drew, a torpedo-plane pilot "willed" to me the canvas and oil paints that he had used for relaxation during off-duty hours. When I had a chance, I made paintings with these materials as Christmas presents for Peggy and for Platt-Forbes.

When the new air group came aboard, there was the inevitable turmoil as pilots hunted for their new quarters, and the passageways were piled high with flight bags and duffels. In the midst of all this, I heard a very distinctive giggle and much laughter coming from the stateroom across the passageway from the one I shared with Dutch. Turning to Dutch, I said, "That *has* to be an old friend of mine named Eddie Seiler," and indeed it was. Eddie was a Princeton graduate, with a wonderful ebullient personality and a talent for leadership. He was a lieutenant in Wordell's fighter squadron, VF 44, and an important part of the glue that held it together. His sisters, Jane and Ursula, were old friends from Haverford days, and Jane was the wife of one of my very close friends, Philip L. Ferris, with whom I had played tennis at Haverford and in New York. Eddie and I had an on-the-spot reunion, and we were to see as much of one another as time permitted aboard ship.

Not so pleasant was the prospect of serving under Captain Wegforth. Few of the officers seemed to know anything about him, but I clearly recalled his fearsome reputation as a "sundowner" when I was stationed on Ford Island, where he was in command of the Naval Air Station. Since no good could come of my sharing my misgivings, I decided to adopt a wait-and-see attitude and hope for the best. As it turned out, it was soon obvious that we were now under a very different kind of commander.

During brief pauses in the action, such as this one in Ulithi and earlier ones at Eniwetok, Majuro, and Espiritu Santo, Dutch and I had organized a poker group which met in the evenings whenever possible to play for modest stakes, far below those in the ready rooms. We both preferred poker to the almost continuous acey-deucy games that were played in the wardroom off-watch. Once in a while someone would have the heel of a bottle left and would share it with the other players. Often the ship's chaplain would join the game and we would make much of how he learned to play poker in Divinity School and how he should contribute half of each pot he won to the ship's welfare fund. Often, too, Joe Wagner, one of the ship's doctors, would sit in, and we would tell him that the price of admission was a bottle of medicinal brandy. These poker sessions in our stateroom were a welcome relief from tension and we all enjoyed them. After the first one in which I won a little money, I decided to see how long I could subsist on poker winnings without drawing any of my monthly pay. Since we left Pearl Harbor, our only use for money was to pay our monthly mess bills and to have a little left over for toiletries and cigarettes (at ten cents a pack, tax free). My good luck continued, and for a year and a half I left all my pay that I had not allotted to Peggy on the ship's books, waiting for me to take out in a lump sum when and if I ever was transferred.

The decision to land MacArthur and his army at Leyte Gulf had now been made and was beginning to be implemented. It was obvious that the enemy would throw against us the strongest force it could muster, and the fast carriers were ordered to hit his main staging areas for bringing aircraft into the Philippines, and try to neutralize them before the landings on Leyte. We had already worked over Luzon and the islands farther south. Now it was time to go north to Okinawa and Formosa (now Taiwan), into the heart of Japan's lines of communications with the Philippines. On 6 October we left Ulithi for an operation that culminated in the greatest battle in naval history.

Sixteen carriers strong, Task Force 38 now could launch one thousand airplanes. Still in Task Group 38.3 under Rear Admiral Sherman, we were again operating with the *Essex, Lexington,* and *Princeton.* With our new air group embarked, we worked hard to accustom the new pilots to the doctrines and procedures of the task force, and found them eager to absorb all the detail we could give them. From the radar standpoint, we knew about what to expect from the other carriers in the group. The *Princeton* had excellent radar and a finely honed and exceptionally efficient combat information center. The *Essex*, flagship of the group, had much less effective radar despite her higher antennae, and less effective fighter direction, perhaps because she was burdened with the extra duty of serving as group fighter director. The *Lexington* was much better than the *Essex*, perhaps on a par with us, and a notch or so below the operating level of the *Princeton.*

Such differences in the effectiveness of radar operations between ships of different classes and between different ships of the same class seemed a curious phenomenon. One would think that, given the same type of equipment, the higher the antennae the greater the range at which a target could first be spotted. Yet, in this respect, certain of the CVLs consistently outperformed CVs with higher antennae. I cannot remember a single time when one of the battleships was first to report a bogey contact. Perhaps the difference was principally a matter of maintenance of the gear and the training and dedication of the radar-set operators and their supervising officers in the combat information center.

Steaming far to the north and closer to Japan than the fast carriers had ever before operated, we launched the first strikes on Okinawa on 10 October. More than 100 enemy aircraft were destroyed, some in the air, many on the ground, and a large number of small warships were caught in the waters around the island and sunk. While 21 of our carrier planes were shot down, most were able to make water landings and many crews were rescued by the hard-working submarines.

From 12 to 14 October, the attack shifted to the airfields on Formosa, where the Japanese counterattacked in great force. The high command in Tokyo ordered all available land-based air to attack and destroy the fast carriers off Formosa, and ordered 300 of their own carrier pilots, as yet only partially trained to replace those lost in the battle off Guam, to assist in the coup de grâce. On the first day of the American attacks, the Japanese launched 101 planes at our carriers, the second day 32, the third 419, and the fourth 199.* After each of their attacks on the task force, the enemy pilots apparently reported staggeringly exaggerated claims of having sunk carrier after carrier, when no such thing had taken place. With each report of spectacular success, more Japanese planes were committed. Eventually, more than 500 of them were destroyed by our fighters and bombers. Incredibly, the enemy reported to Tokyo that it had sunk or damaged 53 of our ships, including 16 carriers! Tokyo Rose announced that the *Langley* had been sunk! Actually, the worst damage to our ships were hits on the heavy cruiser *Canberra* and the light cruiser *Houston*, both of which were severely crippled and taken under tow. To help protect them on the long, slow journey back to Ulithi, Halsey assigned the *Cabot* and the *Cowpens*, two of our sister ships, to provide air cover. We in fighter direction worked at top speed, night and day, for four days with spectacular success. The *Cabot* and the *Cowpens* operated with great skill, considering the relatively few fighters they had on board and the slow speed that the tows forced them to maintain. Both the crippled cruisers reached Ulithi, despite repeated attempts to finish them off. Years later I discovered that the fighter director officer in the *Cabot* was Henry Lowe, another advertising man from New York.

---

*Reynolds, p. 260.

With the landings on Leyte scheduled for 20 October, we now refueled at sea and hurried back to the Philippines for a continuous series of strikes on the airfields of Luzon. After only two weeks on board, Air Group 44 had already had a severe testing, and there was much more to come.

XIV
The Battle for Leyte Gulf

In the opening paragraph of his description of this battle, Hanson Baldwin calls it "the greatest sea fight in history"; he points out that it "sprawled across an area of almost 500,000 square miles," was "as decisive as Salamis," and "dwarfed the Battle of Jutland in distances, tonnages, casualties."* It involved ship-based and shore-based aircraft and every type of warship from submarines and torpedo boats to battleships and aircraft carriers on both sides. It marked the debut of the kamikaze, and the demise of the Japanese fleet. Yet it was fought by American sailors who were very near the brink of exhaustion from an unprecedented ten months of almost continuous operations at sea against the enemy. As Admiral Mitscher put it: "Probably 10,000 men have never put a foot on shore during this period of ten months. No other force in the world has been subjected to such a period of constant operation without rest or rehabilitation."**

The landings on Leyte and the return of MacArthur were called Operation King II. The Japanese plan to oppose them and to wipe out the American naval forces in the process was dubbed the Sho Plan, sho meaning "victory." The latter called for a three-pronged assault on the American forces once they were committed to a given landing area. To the three prongs the Japanese assigned virtually all their warships that were then combat-ready. One prong contained what was left of their carrier forces, the *Zuikaku, Chitose, Chiyoda,* and *Zuiho,* the air groups of which were understrength because of the decimation of their pilots in recent actions, escorted by the *Ise* and *Hyuga,* two hybrid battleships

---

*Hanson W. Baldwin, *Sea Fights and Shipwrecks* (Garden City, NY: Hanover House, 1955), p. 134.
**C. Vann Woodward, *The Battle for Leyte Gulf* (New York: Macmillan, 1947), p. 43.

with very short flight decks added, three cruisers, and ten destroyers. Under the veteran Admiral Ozawa, this prong was to steam south from Japan, approach the northeast coast of Luzon off Cape Engaño and act as a bait to draw Task Force 38 away from Leyte Gulf, where lay the invasion fleet of some 700 vessels, including troop transports and supply ships. This carrier group was designated the Northern Force.

The second and third prongs were made up of surface combatants based near Singapore. The Sho Plan called for them to proceed north to Borneo, refuel at Brunei Bay, and then be divided into two groups. In the dark of night, the Central Force, consisting of five battleships, ten heavy cruisers, two light cruisers, and fifteen destroyers under the command of Vice Admiral Takeo Kurita, would pass through San Bernardino Strait, between the southern tip of Luzon and Samar, thus gaining access to the Philippine Sea and the prizes to be had in Leyte Gulf. At the same time, the Southern Force under Vice Admirals Nishimura and Shima was to take its two battleships, four cruisers, and eight destroyers through Surigao Strait, between Leyte and Dinagat, to join up with the Central Force in the morning and destroy the invasion forces at Leyte.

On the American side, Task Force 38 was not the only combatant force available. The invasion force of transports and supply ships had its own naval escort in the Seventh Fleet under Vice Admiral Thomas C. Kinkaid. With its small escort carriers, CVEs, a group of old battleships, including some rescued from the bottom of Pearl Harbor, cruisers, and destroyers, this fleet was not under Halsey's command. It reported directly to General MacArthur. While Halsey and Kinkaid were supposed to coordinate plans with one another, they had no specific directives and no superior naval officer in the chain of command, short of Admiral Nimitz far away in Hawaii. In the fog of battle, the looseness of the arrangement caused great trouble and misunderstanding.

As in the case of the Battle of the Philippine Sea, the first warning that the enemy was approaching came from the trusty submarine force scouting far to the west. The *Darter* and the *Dace* were patrolling the southern entrance to Palawan Passage, just west of the Philippines, when they obtained a radar contact on Kurita's force shortly after midnight, 22-23 October. Between them, the two submarines sank Kurita's flagship, the *Atago,* and another heavy cruiser, and severely damaged a third. Halsey and Kinkaid were alerted by the submarines' reports and it was estimated that Kurita's force would be within striking range of TF 38's planes by the morning of the twenty-fourth. No intelligence had yet been received on the whereabouts of other enemy forces.

At dawn on the twenty-third, Halsey refueled three of his task groups, while the fourth, Admiral McCain's 38.1, was headed toward Ulithi for rest and replenishment. At the end of the day, he dispatched the three remaining groups to range independently off the Philippine coast, from Luzon in the north down to Samar in the south. Our group was assigned to cover Luzon, and we made our approach close to the Polillo Islands, off the east coast. All night long we were shadowed by Japanese search planes, and there was little time for sleep in radar plot where we spent most of the night tracking bogeys for the night fighters.

For us in Admiral Sherman's TG 38.3, 24 October was one of the most hectic days of the war. Along with the other carrier groups to the south of us, we launched long-range searches at dawn and, by 0830, both the enemy's Central Force and his Southern Force had been spotted. Halsey immediately ordered the three groups to concentrate off San Bernardino and to launch attacks on the enemy ships as soon as within range. He also ordered the far-distant McCain to turn around and come back from his approach to Ulithi. Kinkaid, in turn, ordered his battleship commander to make plans for denying the enemy's Southern Force passage through Surigao Strait.

Before we could launch our deckload strike against the reported positions of the enemy ships, his land-based planes from Luzon attacked us in strength. I suppose that when they found we were only four carriers and were close inshore, they saw it as a great opportunity. At any rate, they sent three waves of attack planes at us in rapid succession, about forty in each of the first two, and even more in the third. Our strike had to be postponed while we borrowed its escorting fighters to augment our combat air patrol and fight for our lives. Our radar operators on the *Langley* did a superb job of detecting and tracking these clouds of enemy airplanes. As we nearly always had the earliest and best information, we were given the assignment of directing the fighters from the other carriers as well as those from our own. Several times during the morning, I directed two interceptions at once. Commander David McCambell of the *Essex* did an astounding job in the ensuing melee, and shot down nine planes before he landed on our flight deck, too short of gas to reach his own. Eddie Seiler told me later that when he first sighted a raid I had sent him out to intercept, the first thing he saw was an exploding Zero, hit by another fighter.

A little after 0930 one of the Japanese dive-bombers came down through low cloud and dropped a 550-pound bomb on the *Princeton*. It went clear through her flight deck and exploded between her main and second decks, setting fire to six fully loaded torpedo bombers on her hangar deck. Her captain, W.H. Buracker, headed her up into the wind, the better to fight the fire, and Sherman detached a cruiser and three destroyers to stay with her and provide antiaircraft protection for any subsequent attacks. Just after 1000, a huge explosion in her hangar split her flight deck open, shot her aft elevator up into the air, and sent black smoke high up above her. In no time at all another raid developed, intent on finishing her off. We intercepted this one with a division of our own fighters which shot down four torpedo planes that were about to launch at the *Princeton*. Our own ship's guns destroyed two dive-bombers from the same group, and the cruiser *Reno* got two more. No more attacks reached the *Princeton* during the morning.

Strikes from the other carrier groups went off on schedule, and throughout the morning were reporting successful attacks on Kurita's ships. Toward mid-morning, our first strike got off from the *Essex* and the *Lexington*, and a second was launched sometime after noon. At 1155 Sherman was ordered by Mitscher to launch searches to the north in an effort to find the enemy's carriers. Just as these search planes were about to be launched, at 1245, our radar in the *Langley* picked up a large incoming raid from the northeast, at a range of 105 miles. We reported it to the flagship and were instantly ordered to direct the four *Langley* fighters that were among the twelve overhead to intercept. I immediately vectored out the *Langley*'s fighters and ordered them to climb to 22,000 feet. Carl Brunmeyer was the leader of this division, and I told him that it looked like a large group of bogeys. After he was well on his way, we asked for and received permission to send eight more fighters from the *Essex*'s contribution to the CAP. The instant Dutch received permission from the flagship, I got them going on the same vector, 035 degrees, and they streaked out after Carl, some ten miles astern of him. When Carl first reported his "tally ho," about forty-five miles northeast of us, there were some sixty planes in the incoming raid, and my heart sank when I had to order him, with only four planes, to "sail into them." I was able, however, to add, "Help is on the way, coming up close behind you."

First Carl's and then the *Essex*'s planes, despite being heavily outnumbered, did a superb job of completely dispersing this raid. At least half were shot down and not one came within gunfire range of the ships. Those that survived turned tail and headed for Luzon. Now the news we had been waiting for came in: the enemy planes had tail hooks, which meant that they had come at us from carriers. This information told Halsey that the enemy carriers were to the northeast, in the area we had still not been able to search.

While that fight was taking place, we detected a new raid coming in: it was ninety miles away and also coming from the northeast. When we reported this one to the *Essex*, the group fighter director assigned its interception to the fighter director officer in the *Essex*, and ordered us all to launch whatever fighters we had left to furnish a new combat air patrol. Unfortunately, because of heavy clouds, the interception was missed, and contact was not made until the raid was within fifteen miles of the ships. Nine were shot down by fighters from the *Lexington*, but an almost equal number of dive-bombers came out of the clouds and dove on the carriers. Many on board thought they were suicide planes, and they may well have been, because we had been warned by intelligence to expect such planes. The barrage of antiaircraft fire that erupted from every ship in the task group brought several of them down while others scored near misses on the *Essex*, the *Lexington*, and on us. Our damage was minor, and we all recovered our returning first strike during a lull in the action around two o'clock in the afternoon.

All during the day, heroic actions to save the *Princeton* were under way. As her fires worsened during the morning, Captain Buracker ordered all but a skeleton crew to abandon ship, and destroyers were kept busy picking the men out of the sea even while the raids were going on overhead. Other ships, including the cruiser *Birmingham* and the destroyer *Morrison*, came alongside to play their fire hoses into the fiercely burning carrier. The *Birmingham* made fast to the windward side of the ship and got fourteen streams of water going into the flames. A volunteer fire-fighting party from the cruiser under Lieutenant Alan Reed went over to the carrier to help the shorthanded crew. The *Morrison* approached on the starboard and leeward side of the carrier, where she was in great jeopardy, and had been pinned there for an hour when her superstructure rolled against the stacks of the *Princeton* and held her against the side as if in a vise. As the afternoon wore on, it was decided that the *Birmingham* would take the *Princeton* in tow. A little before four o'clock, when she had just made fast a spring line to do so, a massive explosion blew off the whole after section of the carrier, raining metal across the deck of the *Birmingham*. Crowded as the deck was with line-handling and fire-fighting parties, 229 men were killed instantly and 420 wounded. Meantime, the air attacks from Luzon's scores of airfields resumed. While we were engaged in intercepting one, a destroyer in the screen reported a sound contact on a submarine. I turned to Dutch, who was bending over the plan position indicator scope, and said something like, "Man, I just can't think of anything else we need!"

We were supposed to be joining up with the other task groups off San Bernardino, to the south, but we could not leave the *Princeton*. We were supposed to be flying searches to the north, but could not spare fighters to escort the torpedo planes to do it. Eventually, we sent them off without escorts. Toward the end of the afternoon, they sighted the Northern Force, which consisted of four carriers accompanied by battleships, cruisers, and destroyers. They were north and east of the northern tip of Luzon, about on the bearing from which carrier planes had attacked us earlier in the day. Although the reports from the search planes were conflicting and somewhat garbled, Halsey now knew where the enemy carriers were.

At a quarter before five, it was decided to abandon and sink the *Princeton* so as to free us for action against the carriers to the north. After all survivors had been taken off, the light cruiser *Reno* sank her with a torpedo and her agony was over. During the day she had lost 108 men killed and 190 wounded. Awful as that was, it was a smaller price than her Good Samaritan, the *Birmingham*, had to pay. It is hard to describe the sadness we felt. The loss of any ship at sea is sad; but the loss of a sister ship, and one with which we had been through so much for so long, was felt personally and intensely. All of us in the *Langley* could visualize ourselves in that fire, in that water. And in CIC we felt keenly the loss of her able fighter direction crew.

Throughout the long day, glowing reports of successful strikes on Kurita's force were being sent to Halsey. Those from the two strikes of our beleaguered task group were much less sanguine than those of the two groups to the south, which were grossly optimistic and highly exaggerated. Because Halsey was embarked in the *New Jersey*, off San Bernardino, and Mitscher was in the *Lexington* far to the north with us, many of these reports went directly to Halsey, without having been screened, questioned, and evaluated by Mitscher's staff. Consequently, they badly misled the Commander, Third Fleet, into believing first, that Kurita had been turned back in the Sibuyan Sea and was retiring from the fight; and second, when information was received that he had again reversed course and was only forty miles from San Bernardino, that the "remnants" of his force had been so badly mauled that they were not a serious threat. In actual fact, only the huge battleship *Musashi*, one of the two largest in the world, had been sunk, and one heavy cruiser so damaged that it had to limp off the stage. His main force was still very much in being.

At Leyte, Kinkaid ordered Rear Admiral Jesse B. Oldendorf, in command of the old battleships of the Seventh Fleet, to prepare to meet the enemy in Surigao Strait. How well he and his chief of staff, Captain Richard W. (Rafe) Bates, carried out this assignment will be described later.

By mid-afternoon, Halsey had notified his task group commanders that four new battleships and five cruisers of Task Force 38 would disengage from their groups and form Task Force 34 under Vice Admiral Willis A. Lee. Kinkaid intercepted this message and assumed that the new task force was then being formed and positioned to guard San Bernardino Strait, just as his battleships would guard Surigao. Unfortunately, he did not intercept a later dispatch from Halsey, sent about five in the afternoon, stating that Task Force 34 would be formed "when directed," indicating that it had not yet been formed. As darkness began to fall, Halsey decided to regroup his three carrier task groups, go north, and, at dawn the next day, attack the enemy's carriers, which had always been his primary target at sea. He missed his chance at them off Midway because he was hospitalized and the command had passed to Spruance, who fought a successful battle but failed to pursue the retiring enemy to the point of annihilation. He missed his chance again off Guam because it was Spruance's turn to command the fleet. Now it was *his* turn. He knew approximately where the Japanese carriers were, and he was going to wipe them out. A little after eight in the evening, he issued the orders. Sherman was to take our group southeast to rendezvous around midnight with Davison's and Bogan's groups, which were steaming north at high speed. Once reunited, we were to prepare for predawn launches to seek and destroy the enemy carrier force.

Having thus got the operation started, Halsey then sent a dispatch to Kinkaid notifying him that he was "going north with three groups" to hit the enemy. To Kinkaid, the message was consistent with his belief that Halsey had already formed a fourth group, TF 34, which would stay behind to guard San Bernardino. Confident that Halsey was watching San Bernardino with a mighty battleship force while the fast carriers were speeding north, Kinkaid proceeded with his own plans. Actually, *no one* was guarding San Bernardino that momentous night.

With our own plan of action now clear, and with no bogeys on the radar screen for the first time that day, Dutch and I had a chance to relax and compare notes. How superbly our radar crew in the *Langley* had performed! How worthwhile it had been to select them so carefully way back at the Philadelphia Navy Yard. How sad it was to lose the *Princeton*, but what a price we had made the enemy pay for her: 120 planes shot down around us and 47 more over Luzon! And now what a chance to get the carriers. Exhausted as we were, we both recognized that it had been a day we would never, could never, forget. "Dutch," I said, "we truly had an angel on the yardarm all day today, and I sure hope she stays there tomorrow."

With that, I went below to get three hours of sleep before taking over the midwatch a little before midnight.

When I took over the watch in CIC, we were just converging at 25 knots on the other two task groups. From on deck, not a solitary light could be seen on the horizon; yet on the radarscopes all of Bogan's and all of Davison's ships were racing northward directly in our path. It was not only thrilling to see us swing into formation alongside them, taking station by radar, but it was a great comfort to have friends around us again. Soon afterward, Sherman detached the ships that had been damaged in trying to save the *Princeton*: the cruiser *Birmingham* and the destroyers *Gatling, Irwin*, and *Morrison*. They dropped out of the formation and headed for Ulithi and repairs. A few hours later, Halsey ordered Task Force 34 to form up, and one by one they, too, dropped out of formation and regrouped as an entity of their own composed of six battleships, seven cruisers, and seventeen destroyers.

Our hangar deck crew worked strenuously most of the night to rearm and refuel all our airplanes, preparing for a strike very early in the morning. Heavy belts of fifty-calibre ammunition were fed into the fighters for each of the six guns buried in their wings; and bombs and torpedoes were trundled out for loading into the torpedo planes.

Unbeknown to us, as we concentrated on our plans, Kurita was even then taking his entire force through San Bernardino Strait unopposed and unobserved by either Halsey's or Kinkaid's commands, and Nishimura was heading into the trap that Oldendorf had carefully laid for him in Surigao Strait. Having positioned torpedo boats at the lower end of the strait, Oldendorf was able to get early warning of the Japanese approach and to launch effective attacks by torpedo boats and destroyers as the enemy proceeded up the strait in column formation. Waiting at the top of the T, Oldendorf's old battleships and cruisers were ideally situated to cross the T and pour broadside after broadside into the Japanese column. This was undoubtedly the most successful night surface action of the war. None of Nishimura's or Shima's ships got through. Most of Nishimura's heavy units were sunk or badly damaged; most of Shima's were routed and retired. American losses were minimal.

Kurita fared far better. Undetected until he suddenly appeared after daylight within shelling distance of Kinkaid's thin-skinned escort carriers, he engaged in a wild melee with the planes they were able to launch and with the destroyers and destroyer escorts of the Seventh Fleet, the only forces between him and the transports in Leyte Gulf. These little ships attacked so heroically that Kurita became convinced he was up against units of Halsey's force. During the melee, suicide planes of the newly formed Kamikaze Special Attack Force made their first appearance. Flying from fields on Luzon, they caused severe damage to four of the CVEs and succeeded in sinking one, the *St. Lo*. Another CVE, the *Gambier Bay* was sunk by shellfire, and two destroyers and a destroyer escort were sunk. Seven CVEs in all were damaged. Serious as these losses were, they were modest in comparison with the castastrophe that would have taken place if Kurita had pressed on into the gulf.

The moment Kurita appeared on the horizon, Kinkaid began sending frantic messages to Halsey asking for immediate help and trying to find out the position of Task Force 34. By ten in the morning, even Nimitz had sent a dispatch to Halsey asking where TF 34 was; as he, like Kinkaid, assumed that it was guarding San Bernardino. Not until then, and much too late, was drastic action taken to send help to Kinkaid.

Meanwhile, we had been steaming north all night. TF 34 was ranged some ten miles ahead of us to engage the enemy's surface ships when we caught up to them. At first light in the morning of 25 October, we launched not only search planes but also a strong strike force. The strike was to orbit about fifty miles ahead until the search planes pinpointed the enemy position, and then hit the enemy carriers before they could hit ours. At 0735, when the planes had been up for an hour and a half, the enemy force was sighted and the strike force vectored over to intercept it, 140 miles northeast of us. Although we didn't know it, the four carriers in Ozawa's force had been so severely stripped of their airplanes that, among them, they could launch only about twenty. Ten days earlier, 150 of them had been taken away to operate against us from bases in Formosa; and Ozawa had committed 80 others in his strike against us the day before, off Luzon. Clearly, the only reason he pressed on when he had only 20 planes left was to act as bait to draw TF 38 away from Leyte Gulf.

Our first strike hit the enemy at 0840 and quickly dispatched his defensive fighters in the air. The lack of planes on the flight decks of the carriers beneath them puzzled our pilots and, as time wore on, we were puzzled that no retaliatory strike was launched at us. Our dive-bombers and torpedo planes went to work with a vengeance, sinking one carrier and scoring hits on other carriers and a battleship.

We launched a smaller second strike, which arrived over the target at 1010 and again hit the three remaining carriers, a battleship, and a cruiser. The enemy ships then retired northward; the damaged ones fell behind and some, dead in the water, became sitting targets for the big guns of TF 34, which were coming up fast. It looked as if our battle plan was working perfectly.

At 1000, however, the whole situation had changed. Faced with a fifth frantic appeal for help from Kinkaid and with Nimitz' query about TF 34, Halsey had to change his plans radically. He had ordered McCain's task group to join the other carriers as soon as it had finished refueling. Now, he countermanded that order and sent McCain at flank speed to the rescue of Kinkaid, hoping that he would arrive in time for his planes to attack Kurita's force. Simultaneously, he ordered TF 34, almost in sight of Ozawa's cripples, to reverse course and dash all the way back to Samar, an almost hopeless chase after Kurita. This latter order meant that the battleship sailors, who had waited a lifetime for the opportunity that lay just over the horizon, were to be denied; it also meant that the further destruction of Ozawa's forces was now entirely up to TF 38. During the day, we launched three more strikes, making five in all, but we were handicapped by the wind being from the east, which required us to turn away from the enemy every time we launched or landed planes, while Ozawa was escaping to the north.

In the course of our five attacks, all four of the Japanese carriers were sunk, along with one of their cruisers and two of their destroyers. One of the latter was heavily loaded with survivors from other ships, and, on the last strike of the day, under Mac Wordell our fighters sank it by strafing. The cruiser *Tama* was also sunk, but the two battleships and remaining cruisers were able to retire despite their damage. Without much doubt, they would all have been sunk if TF 34 had been given just a few more hours on the job. The four carriers we sank were the *Zuikaku*, veteran of the Pearl Harbor raid, the *Chitose*, the *Chiyoda*, and the *Zuiho*.

With our planes back on board by dark, the carriers were ordered to break off and return to the vicinity of Leyte, where for the next few days we were to resume air support for MacArthur. On the twenty-eighth we were ordered back to Ulithi to rearm, refuel, reprovision, and get some sleep. In the entire three days of the Battle for Leyte Gulf, the Japanese lost 305,710 tons of warships: four carriers, three battleships, ten cruisers, and nine destroyers; we lost the *Princeton*, two escort carriers, two destroyers, and one destroyer escort, amounting to 36,600 tons. Both sides lost hundreds of airplanes. For years afterward, Halsey was criticized for having "left the gate open" at San Bernardino, but the criticisms never seemed to take into account what might have happened later on if the enemy had had those four carriers at Lingayen Gulf, at Iwo Jima or off Okinawa.

Leaving Bogan's and Davison's task groups off Leyte, we returned with McCain's group and reached Ulithi on 30 October.

XV
Supporting the Troops on Leyte

After only two days devoted to rearming and reprovisioning, we were recalled to render further support to MacArthur's troops on Leyte. Torrential rains ever since his landing had made it virtually impossible to activate the primitive airfields taken from the Japanese. Only Tacloban was operational, and it had too few Army Air Force fighters to maintain an effective combat air patrol and attack the convoys that frequently reinforced the Japanese troops on Leyte. Until this situation could be corrected, it was up to TF 38 to pinch hit.

While we were steaming toward Ulithi, the carriers *Franklin, Belleau Wood* and *Intrepid*, "holding the fort" until our return, were all hit by suicide planes. On 1 November the enemy made a strong counterattack on MacArthur, and it became our mission to suppress enemy air strength on his reinforced fields on Luzon. On the night of 3 November, when we were en route to our launching position, an enemy submarine somehow penetrated our screen of destroyers and torpedoed the cruiser *Reno* which was close alongside us. She lost steering control and was taking on water fast. After damage-control parties had fought hard to save her, Admiral Sherman detached four destroyers to escort her back to Ulithi.

Admiral McCain had by this time relieved Admiral Mitscher as task force commander, but Halsey still commanded the fleet. With three full groups off Polillo, where we had lost the *Princeton*, we had far more strength. On 5 and 6 November, we found a great many airplanes on and around the Luzon fields, and in the course of heavy strikes destroyed some 400, most of them on the ground. In retaliation, kamikazes struck, and one of them eluded the CAP and hit the island of the *Lexington*.

On 11 November, we sighted a Japanese convoy en route to Leyte. It consisted of five transports carrying 10,000 troops and was escorted by seven destroyers. When it had almost reached the landing area in Ormoc Bay, we attacked it with a massive strike of 347 planes from the three task groups. Although some 30 enemy aircraft tried to interfere, our planes dove on the transports and sank all five before the troops could land. Then, having shot down half the enemy planes and dispersed the rest, they concentrated on the destroyers, sinking four on the spot and two that were trying to return to Manila Bay. What a way to celebrate Armistice Day!

Returning to our launching area off Luzon, we next caught another large group of ships in Manila Bay, and on the thirteenth and fourteenth sank a light cruiser, five destroyers, and seven merchant ships. During the action, fighter sweeps over nearby Clark Field destroyed an estimated seventy-five planes. After one more strike, on the eighteenth, our group was sent to Ulithi to pick up where we had left off when we were so suddenly recalled on 1 November.

On 20 November, we were at anchor in the lagoon at Ulithi, getting ready for our next sortie. At dawn on that windless morning, the sea in the lagoon was glassy smooth. Suddenly, with no warning whatever, our lookouts sighted torpedoes slicing through the translucent water. One went right past our bow, another hit an anchored tanker. The entire lagoon awoke to frenzied activity. Ships went to general quarters. Destroyers got under way to find the culprit. Radar antennae began to rotate and radios crackled. In radar plot we could make little sense out of the situation, and it was only long afterward that we learned the attacks had come from midget submarines, which had somehow sneaked through the patrolling destroyers and the torpedo nets at the entrance of the lagoon. I was on deck and could clearly see clouds of black smoke going straight up in the air from the tanker *Mississinewa*: the fire had started in her cargo of 400,000 gallons of aviation gas. With no wind, there was no windward side from which her crew could escape and some fifty officers and men perished. Our destroyers sank two of the midget submarines.

By this time, it was obvious to all concerned that Task Force 38 must somehow change its defensive tactics, if the carriers were to survive the onslaught of kamikazes. A message from Sherman invited all hands to offer suggestions as to what might be done. Recalling my experience in the destroyer *Bennett* while I was in radar school at Pearl Harbor, it occurred to me that by placing a fighter director on board certain destroyers we could use them as radar pickets. If they were deployed on the enemy's line of bearing but well in advance of the carriers, they ought to be able to detect the kamikazes before we could, give us early warning, and even direct fighters to intercept that much farther away from the carriers. After a few tries, I worked out a disposition on a maneuvering board, wrote up a rationale for the idea, and showed it to Commander Hannegan, who by this time had been promoted to executive officer under Captain Wegforth. Hannegan studied my proposal with interest and approved its being forwarded to Admiral Sherman. I am certain that many others must have had the same idea, and claim no credit for having originated it, but the fact is that shortly thereafter, radar picket destroyers became our way of life, along with heavily increased combat air patrols and enlarged complements of fighters in the *Essex*-class carriers. My submission was never acknowledged, nor was my offer to volunteer as the fighter director officer in one of the destroyers to try out the idea.

Important changes in the command structure of the *Langley* were now in place. Commander William Guthrie had become our air officer, and several of our most senior lieutenants, including Dutch Doughty, Tom Smith, McKee Thompson, and Glen Butler, had been promoted to lieutenant commander. Meanwhile, it became common knowledge in the wardroom that Hannegan and Guthrie were taking heavy criticism from Captain Wegforth for deficiencies he perceived in the ship he had inherited from Dillon; and other ship's officers were feeling the heat in their own areas of lesser responsibility. Sensing a sea-change in the climate of the wardroom, I concluded that its cause was simple: Dillon had *led*, Wegforth *drove*.

About this time we lost Dutch to the *Essex*. It came about suddenly. One of the senior fighter directors on the admiral's staff went ashore in Ulithi, to Mog Mog Island, which had been designated a recreation area for swimming and having a beer or two, or three. This officer had apparently had quite a few by the time he returned to the flagship, where he fell down an elevator well and broke a leg. In no time at all, dispatch orders arrived in the *Langley* for Dutch to be detached and to report to the *Essex* as task group fighter director officer for TG 38.3. We had been roommates and close friends for more than a year, and I was very sad to see him go. We gave him as good a sendoff as we could and consoled ourselves with the notion that at least we would now have a friend at court on Admiral Sherman's staff.

Gus Rounsaville having been detached some months earlier, this development left me as the senior officer in radar plot and Commander Hannegan ordered me to take over responsibility for it.

Our final action in support of the Leyte landings took place on 25 November. By this time, the Army had enough planes and airport facilities to take over the defense of Leyte. Our search planes having sighted a number of ships in and around Lingayen Gulf, on the west coast of Luzon, we sent strikes clear across the island to get them. On the way, the escorting fighters encountered many enemy planes in the air and shot down twenty-six of them. Of the Japanese ships, the strikes were able to sink a cruiser, a frigate, and five ships about the size of our LSTs. Unfortunately, however, we paid a big price. The kamikazes came out in force all day long, and we were as busy in CIC as we had been on 24 October. When we cleared the plotting board at the end of the day, we were up to Raid 27, one of the highest numbers I can remember in a single day. While most were intercepted successfully, those that got through hit the *Intrepid*, the *Essex*, and the *Cabot*, again underscoring the menace of the kamikazes. The importance given to intercepting even a single aircraft under these conditions is well illustrated by the following quotation from the *Langley*'s Action Report for the Lingayen Gulf Shipping Strike, 25 November 1944:

"The interception described below is believed notewothy as an example of the possibilities of long-range fighter direction with present radar and radio equipment:

"A single bogey 80 miles to the northwest was detected by the ship's SK radar. This bogey was on course 170 degrees True and was therefore not an immediate menace, but one division of *Essex* fighters under *Langley* control was orbited 30 miles west of the Task Group as a precaution in case the target changed course and started to close. When it became apparent that the target was continuing in a southerly direction, the fighters were given a westerly vector and a tally-ho was obtained at a point approximately 260 degrees True 70 miles from the ship. The fighters lost visual contact with the enemy almost immediately, however, and their target changed course sharply to the southwest and then to the west. The fighters were vectored accordingly, and tally-hoed again at 250 degrees True 92 miles from the ship, at which point both friendly and bogey indications faded from the radar as the fighters came down from 12,000 feet to make their runs. Before radio contact was also lost the fighters were given a homing course. After approximately ten minutes radio contact was regained with the fighters who were then following their homing vector having shot down one 'Frances' well inland of the Luzon coast. Radio communications were considered excellent on channel A, and the fighters returned safely without further incident."

Officers and enlisted men of the *Langley*'s radar plot seated in front of the ship's scoreboard.

XVI
The Task Force Meets a Greater Power

The next scheduled operation before landing on Luzon itself was to be MacArthur's assault on Mindoro, beginning 15 December. Because of the island's close proximity to Luzon, it was particularly important to maintain tight control over the Luzon airfields to avoid interference with the general's assault force and supply ships coming up from Leyte. The Army was now in position to "keep the peace" south of Manila, but everything north of it was to be the responsibility of Task Force 38.

The arrival of new carriers and the need to repair those hit by kamikazes caused another round of changes in the task force. It was to be operated in three groups: we were still in TG 38.3 under Admiral Sherman in the *Essex*, but the other carriers in the group were now to be the *Ticonderoga* under Captain Dixie Kiefer, whom I remembered as having been the resident naval inspector at Pratt & Whitney before the war, and one of our sister ships, the *San Jacinto*. We left Ulithi on 10 December for our station off the east coast of Luzon.

For three very full days we hit the numerous fields around Manila and those farther to the north. While many aircraft rose to meet our fighter sweeps, the majority were caught on the ground, sometimes interspersed with dummies parked in revetments or under the palms alongside the runways. After three days of heavy poundings during which Halsey claimed to have destroyed 270 enemy planes, we withdrew on 16 December to a refueling rendezvous with the tanker fleet next day.

Early the next morning, we met the tankers on the open sea, and began the delicate operation of coming alongside, rigging hoses from tanker to ship, and starting to take on fuel. With the wind steadily rising and visibility steadily lowering, it became more and more difficult to keep station alongside a tanker; and, as ships of various sizes rolled out of synchronism with the oilers, the big hoses began to part and spew oil. When destroyers, dangerously low on fuel and high in the water, tried to refuel in the lee of battleships, they too found it impossible to keep station and there was real danger of collision. The refueling was postponed to the next day and to a new location, at which it was hoped the conditions would be better.

Although it was still not recognized, we were in the path of a furious typhoon. First thought to be a "tropical disturbance," this typhoon was bearing down on us with no more warning than the signs we could see around us. As it became more intense, the crucial question became one of estimating where the storm's center was so that we could lay a course away from it. Each carrier had an aerologist on board and, as early as 0600 on the seventeenth, ours in the *Langley* estimated its position, but missed it by approximately 180 miles. Admiral Halsey, who was in the battleship *New Jersey* and had a very experienced aerologist on his staff, eventually ordered us all to send our best estimates of the storm's center and its course and speed to our task group commander. Apparently, there was no consensus, and Halsey, as fleet commander, changed the rendezvous with the tankers three more times without success. Because of his commitment to MacArthur, he was reluctant to give up the schedule of striking Luzon again on the eighteenth; however, any thought of refueling was out of the question. He released the destroyers lowest on fuel to join the tanker fleet so that they could refuel when conditions abated, and searched for a course that would take his fleet out of danger. It was by then much too late.

In addition to my duties as officer in charge of CIC in the *Langley*, once a week I stood a watch as Air Department duty officer from noon of one day to noon of the next. I was duty officer from noon on 17 December until noon on the eighteenth. When the fueling attempt was abandoned on the seventeenth, all flight operations were necessarily also abandoned, and Commander Hannegan ordered us to prepare for the storm to worsen and to take every precaution. In the afternoon, I left radar plot to the capable watch officers there, and worked for hours with the flight deck and hangar deck officers and crews to secure everything movable. We first lowered the heavy torpedo planes to the hangar deck and lashed them down with double and triple the number of restrainers we usually used. Then we did the same with the fighters on the flight deck for which we could not find room in the hangar. When we ran out of wire cable, we got manila line from the bosun's

Task Force Meets a Greater Power   119

locker and made another round of restrainers, pulling everything as tight as we could possibly get it. Finally, with all thirty-three airplanes "trussed up like Christmas turkeys" we set about lowering to the hangar all the other rolling stock: tractors, bomb-handling dollies, forked lifters, and anything else that wasn't nailed down. Lastly, we lowered the forward elevator and left it down. It was far into the night before we were satisfied that we had the center of gravity of the ship as low as possible, and the gear of the Air Department as well secured as human beings could make it.

By the morning of the eighteenth, visibility had shut down to almost zero. The sky and the sea seemed to merge together into a single plane. The wind increased to 70 knots, still rising. Waves became fifty and sixty feet high. The officer of the deck could no longer see the bow of his ship, much less the neighboring ship on which he was to keep station. For approximately twenty-four hours, all station-keeping was done by radar, and so important were our minute-by-minute reports of the range and bearing of the guide ship that I assigned an officer to help our enlisted talker on the bridge. As the waves increased to, as some said, seventy feet in height, steering became almost impossible for the smaller ships, and ships would call out repeatedly on the TBS circuit to warn that they were out of control. Every time one of these passed helplessly through the task group formation, it was our responsibility in radar plot to track her and, if she was in danger of colliding, to give the officer of the deck an evasion course that would not put us in collision with some other ship. The destroyers in the screen, the smallest ships in the group, were often invisible even on the radar screen, as mountainous seas swept over them and they plunged into the cavernous troughs between waves. We forbade the crew to go out on deck, except for changing the watch along the huge hawsers we had rigged as lifelines the day before. Cooking was out of the question, and the galley made sandwiches around the clock. The CVLs in the task force, our sister ships, began to report that aircraft had broken loose, and either slid overboard from the flight deck, or, even worse, crashed into others on the hangar deck and started fires. The violent rolling of the ship made it almost impossible to stop a loose plane. As report after report came in from the *San Jacinto* in our task group and from the *Monterey*, *Cowpens*, and *Cabot* in others, I kept my fingers crossed and prayed that we had put down enough cables and lines to save our own.

At the height of the storm, we rolled through 70 degrees and dipped green water into the guntubs at the edge of the flight deck. Water

120   Angel on the Yardarm

The *Langley* rolls heavily in the typhoon of December 1944, while a battleship rides steadily astern. Photo from the National Archives. Original taken from the USS *Essex*.

somehow got into the ship's ventilation system, and at one point we had to secure the SK radar because salt water had been forced up an air intake and flooded part of the set. At another point I heard a miracle take place. A cruiser ahead of us in the formation called, "man overboard" on the TBS. Visibility was zero. We could not see the ship, much less the man. Yet, a few minutes later, I heard the ship astern of us come on the TBS with the incredible news, "We have your man!" In all that maelstrom, the bow of the ship astern must have simply spooned him up out of the waves and deposited him on the forecastle!

With little food, no sleep, and great physical exertion, we kept going simply because we had to in order to survive. In the *Langley* we never knew just how hard the wind finally blew for the simple reason that the cups on our anemometer blew off when it reached 105 knots. Our inclinometer went up against its stops. The wind scoured paint off the sides of the ship, and the sea rolled up steel stanchions as big around as my wrist. Yet the airplanes held.

In tragic, lonely battles with the sea, three destroyers were completely overwhelmed. Beyond visual or radar sight of the rest of the force, the *Spence, Hull,* and *Monaghan* capsized and sank. For days afterward other destroyers and search planes were assigned to look for survivors, and eventually ninety-eight were saved. Seven hundred and ninety were not.

When the storm finally passed by and the damage was added up, 146 planes had been lost, nearly all from the other CVLs in our force and from escort carriers with the tanker fleet; eighteen ships required major repairs, and nine suffered lesser damage. It took days to reassemble the scattered units of the fleet, to search for survivors, and finally to refuel. At one point, I heard a message from Halsey to all hands saying, "I will never, repeat *never,* trust an aerologist again!"

After dispatching the damaged ships to Ulithi, Halsey made an abortive attempt to hit Luzon again. The weather off Luzon continued to be so bad that he cancelled strikes and sent us all to Ulithi, badly shaken and exhausted. We in Sherman's group arrived there on the morning of 24 December.

Whole books and many fine articles have been written about the Third Fleet's engagement with this typhoon. Several have included a great photograph of the *Langley*; it was taken from the *Essex* and shows her at the end of a heavy roll. Of all the accounts I have read, none brings back the taste of the salt, the howl of the wind, and the crash of the seas as well as does Hanson W. Baldwin's for *The New York Times*

Magazine of 16 December 1951, entitled "When the Third Fleet Met 'The Great Typhoon.'" Written in an almost biblical style peculiarly appropriate to the subject, its first paragraph began:

"It was the greatest fleet that had ever sailed the seas, and it was fresh from its greatest triumph. But the hand of God was laid upon it and a great wind blew, and it was scattered and broken upon the ocean ... more men lost, more ships sunk and damaged than in many of the engagements of the Pacific War."

When our anchor finally bit into clean coral sand in Ulithi Lagoon, I headed for my bunk with a feeling of pride that the *Langley* had not lost a single man or airplane, but with a feeling of great sadness for all the human losses the fleet had taken. As Samuel Eliot Morison put it, "For some reason that goes deep into the soul of a sailor, he mourns for shipmates lost through the dangers of the sea even more than for those killed by the violence of the enemy."*

Hardly had I reached my stateroom and begun to take off my clothes, when the loudspeaker on our passageway blared out, "Now hear this: Lieutenant Monsarrat, quarterdeck."

Wearily, I put my shirt back on and went down to the quarterdeck to see what the problem was. "John, we've got to get the Christmas mail on board," said the officer of the deck, "and you've been assigned to take an LCVP and get it."

It was afternoon before we could borrow an LCVP; neither of the two little whaleboats we carried was anywhere near large enough to carry sacks containing mail for more than one thousand crew members, consequently it was quite late in the day when I shoved off with a coxswain, a chart showing approximately where the mail barge was, and a working party of a dozen sailors. There being hundreds of ships anchored all over the lagoon, finding the mail barge was a task in itself and it was dark when we finally came alongside. Now, of all things, the crew of the barge told us that the *Langley*'s mail was in the hold underneath seventy-five sacks of *Wasp* mail! I was so tired that at first I was ready to give up. But then the situation began to strike me as funny, and finally, as a huge moon came up to flood the lagoon with light, I began to enjoy the experience. Joining the sailors on my work party, I helped uncover the *Langley* mail and manhandle the sacks over the side into our LCVP. It was good catharsis for all the tragedy we had so recently left. On the way back to the ship, I reflected on how much we

---

*Morison, v. XIII, p. 59.

had to be thankful for, and how that angel must have stayed perched on our yardarm. The sadness began to soften, and the thought of what the mail meant to all those homesick sailors made the excursion worthwhile. By midnight we had it all on board, and at last I could make my way back to my bunk.

XVII
Lingayen Gulf and the South China Sea

Our stay in Ulithi lasted only six days, and it was a particularly hectic time. In addition to the usual rearming and reprovisioning, we had to do our best to repair the damage from the typhoon. Unlike the more severely damaged ships, we did not come alongside a repair ship for assistance. We did most of the work ourselves. Once again Tom Sorber proved to be an absolutely invaluable member of the crew. He diagnosed the saltwater damage to the SK radar, got it dried out, and somehow scrounged enough spare parts for our radarmen to repair it. His inspection of the SK and SC antennae revealed that they were heavily salt-encrusted, and men were hoisted up in boatswains' chairs to give them a thorough cleaning. By the time we were ready to put to sea, our radar was as good as ever; many other carriers were not so fortunate.

While we were involved with radar repairs, welders and riveters were busy replacing stanchions, guntubs, the anemometer, and other items that had been torn loose by the storm; and the smashed crockery in the galley and the wardroom was replaced. It all added up to a tall order to fill in just six days!

Somewhere along the line, Captain Wegforth found time to hold a "quarters for muster, flight deck parade" in order to present decorations that had been received on board from Admiral Halsey in recognition of individual performances during the Battle of the Philippine Sea. Most of these very properly went to the pilots who had done such outstanding work, but Dutch Doughty and I were also singled out for our work in intercepting the big raids off Luzon on 24 October. Dutch, now of course in the *Essex*, received the Legion of Merit. On the flight deck of the *Langley*, I was called forward to be "pinned" by Wegforth with the Naval Commendation Medal, while Commander Hannegan read the citation over the loudspeaker system:

"The Commander, Third Fleet, United States Pacific Fleet, takes pleasure in commending
LIEUTENANT JOHN MONSARRAT
UNITED STATES NAVAL RESERVE
for service as set forth in the following
CITATION:
"For outstanding service in the line of his profession while serving as Intercept Officer of the U.S.S. LANGLEY during the Japanese air attack on the Task Group to which the LANGLEY was attached on 24 October, 1944, in the Philippine Sea. By his demonstration of exceptional skill and outstanding devotion to duty, at least three successful interceptions of air raids were effected, resulting in the routing of enemy forces at a considerable distance from the Task Group, thereby contributing to the ultimate destruction of Japanese air forces in the area in which the fleet engagement occurred. His conduct at all times was in keeping with the highest traditions of the naval service.

W.F. Halsey
Admiral, U.S. Navy"

While I was naturally very pleased at this recognition by the admiral, I felt self-conscious about it and somewhat guilty at being singled out. Like everything else in radar plot, our work on 24 October was strictly a team effort.

Our next operation was scheduled to be in support of MacArthur's return to Luzon through landings in Lingayen Gulf, on the west side of the island. The operation plan called for us to blanket the enemy airfields on Formosa as well as Luzon in advance of the landings on 9 January. A vast attack force under Admiral Kinkaid was to bring the general and his troops up from Leyte through the South China Sea, close to Mindoro, past Manila Bay, and into Lingayen. Escort carriers and Army fighters were to protect them on the way north; our job was to keep planes on Luzon and Formosa from attacking them from the north. This was familiar reading, but I could hardly believe my eyes when I read an addendum that was to be executed if and when Halsey called for it. It directed us to strike Formosa again, then head *west* into the South China Sea and steam clear across it to the coast of French Indochina (now Vietnam) to seek and sink Japanese warships and merchant ships believed to be near Saigon. This foray into what amounted to a Japanese lake would expose us to attack by land-based aircraft from all directions, and it looked to me as if we would have to be very lucky indeed to get away with it. Halsey, however, had an incentive. He had received reports that the two battleships that had escaped from Ozawa's force while we were sinking his carriers, the *Ise* and *Hyuga*, were at Camranh Bay, and he wanted badly to eliminate them from what was left of Japan's "force in being."

To attend to first things first, TF 38 sortied from Ulithi on 30 December and, after refueling at sea along the way, headed for a launching point off the east coast of Formosa. As we steamed north, the weather turned foul and we were hindered by rain, high winds, and heavy cloud cover for almost the entire month of January. We were able, however, to get off strikes against Formosa and as far north as Okinawa on 3 and 4 January. During one of these we lost another fighter pilot from VF 44, Lieutenant Charles August, who was seen to bail out of his damaged plane. Chuck was one of the pilots I came to know in the Bureau of Aeronautics, when he had just returned from being shot down over Africa during the Casablanca landings. He had spent three days in a French jail, knowing all the time that the Navy offshore was about to bombard the docks a few blocks away! Years later, I saw Chuck again in New York and only then learned that he had survived and been rescued from a prison camp at the conclusion of the war.

Part of our assignment was to shuttle back and forth between Formosa and northern Luzon: there was an obvious advantage in keeping the enemy guessing where we would strike next. Accordingly, the planners devised an interesting scheme to deceive the Japanese into thinking we were spending the night close to Formosa, when actually we left at nightfall on a high-speed run to Luzon. This activity entailed the fabrication by each carrier of devices known as "gulls": to four-foot planks about the size of two-by-fours we attached balloons containing some element that served as a good radar reflector. We made up about a dozen of these devices, the balloons being attached to the planks by heavy cord, and the other carriers did the same. Then, as it got dark and the task force headed for Luzon, the order came over the TBS circuit to "stream gulls." Tom Sorber and I were so interested in the idea we streamed the gulls ourselves off the *Langley*'s fantail, then rushed back to CIC to see what we could see. On the SG radar, they could be clearly seen, falling astern in our wake, and, at a distance, might well be taken for ships. An hour or two later, we saw bogeys flying out from Formosa, and had the satisfaction of seeing them orbiting over our gulls in the dark. We hoped they would drop their bombs on the gulls; but in any event they were helping us achieve surprise in the morning on Luzon.

Although we were successful in keeping planes from Luzon and Formosa away from the Lingayen assault force, kamikazes based farther south, in the island chain, inflicted a terrible beating on it. Between 3 and 6 January, no fewer than twenty-five ships were sunk or damaged by almost constant kamikaze attacks. The passage of the assault force up to Lingayen Gulf became one of the riskiest of the war, and we wished that we could have been in two places simultaneously to help them. Although the force had many CVEs with it for protection, its proximity to mountainous islands made detection of enemy planes difficult and most interceptions were at very short range. After a traumatic passage, troops gathered from sixteen bases all over the South and Southwest Pacific were successfully landed on the beach at Lingayen. Here, the Japanese followed the tactic they first used at Palau: they withdrew to interior positions and allowed our troops to come ashore on 9 January with very little opposition.

Halsey was immediately unleashed to put the addendum to the plan into effect. After hitting Formosa again on the ninth, he took Task Force 38 west through the Bashi Channel in the dark of night and under the strictest radio silence, entered the South China Sea undetected. At the same time, he ordered the tankers to go in through the Balintang Channel and be ready to refuel us during the operation. By noon on the eleventh, we had our fuel and proceeded to a position just off the Indochina coast, from which we launched daylight strikes at Camranh Bay. The two Japanese battleships had left sometime earlier and, without our knowledge, had gone south to Lingga Roads, near Singapore. Fortunately, there were plenty of other desirable shipping targets in the harbors and just offshore. In strike after strike, we attacked three large convoys, including one of fifteen ships off Quinhon, against which the *Langley*'s planes were particularly successful. Nine precious tankers and an escorting cruiser in that convoy alone were sunk. In all, fifteen Japanese warships were sunk, as well as twenty-nine merchant ships, including a dozen of the tankers Japan so badly needed. Halsey called it "one of the heaviest blows to Japanese shipping of any day of the war."

Air opposition was scarce. Many planes were destroyed on the ground and the few that took to the air to retaliate were quickly dispatched. One intercept, however, led to a mystery. At a time when Sherman's task group had moved to a position only thirty miles off the beach, we detected a bogey some forty miles inland over the mountain range that faced the coast. Because of the mountains, it was difficult to track the plane and few other ships were able to do so. Since it was heading our way, the *Essex* ordered us to direct fighters from another carrier, which were then flying combat air patrol, to intercept. We vectored them in over the land, and Harry Polinsky, one of our best radar operators, did a superb job of tracking the bogey through the mountains. In due course, the fighters found it and reported that it was an Emily. They shot it down and we all felt that this was an exceptionally good demonstration of our operator's skill. After the CAP landed later in the day, we got a blinker message from the flagship saying that the gun cameras in the fighters had filmed the attack and the bogey was now identified as a Consolidated B-24 in a type of camouflage that was considered obsolete. I had directed the interception and had repeatedly checked to confirm Polinsky's opinion that the bogey never showed IFF, which would have marked it as a friendly plane. We had at times been warned by intelligence that some captured American planes might be used against us, and since there were no American B-24s based anywhere near this area, the warning seemed the best explanation. It remained a mystery, and I wished that the pilots who shot down the plane had come from our ship, in which case we would have been able to talk to them in detail about it.

After sunset we turned away from the coast and headed northeast to another rendezvous with the tankers. A typhoon was in the making, and this time we were on a good course to avoid it, but the fringe winds and seas were getting difficult. After refueling on the fourteenth in heavy weather, the task force ranged north to hit Takao on Formosa, Hainan, the Chinese ports of Amoy, Swatow, Canton, and Hong Kong, successively. In each port, we looked for the elusive battleships, which were actually far to the south, but instead found more tankers and freighters to attack. With the weather steadily worsening, Halsey decided to refuel one more time under the lee of Luzon and then get back out of the "lake" through the Balintang Channel.

During the afternoon and evening of 20 January, we headed northeast through the channel, expecting to be heavily attacked. All we found were frequent flights of planes operating between Formosa and Luzon, presumably evacuating key personnel, and intercepted fifteen of them. We steamed through the night to a launching position off the east coast of Formosa. In ten days we had steamed 3,800 miles through the South China Sea surrounded on every side by Japanese-held territory.

XVIII
The *Langley* Takes Her Lumps

The twenty-first of January was to be a fateful day for Task Group 38.3 and for three of its ships, including the *Langley*. From a position one hundred miles east of Taito, on the southern coast of Formosa, all three task groups launched predawn fighter sweeps over the island in the best weather we had seen in a long while. There was a fleet of tankers and cargo vessels in Takao harbor, and we devoted most of the morning to attacking them; we sank ten, including five tankers. Few airplanes rose to meet our strikes, but many were found on the ground and destroyed by subsequent strikes throughout the day. In all, some 104 were claimed.

It was at sea, however, that our main defensive action took place. The three task groups were ranged on a northwest-southeast line, Sherman's group being farthest to the north, Bogan's TG 38.2 in the middle, and Admiral Arthur W. Radford's 38.1 farthest to the south. Toward the end of the morning, there having been no enemy air activity over 38.3, our two battleships, the *Washington* and the *North Carolina*, took the opportunity to fuel destroyers alongside. To facilitate this operation, the group steadied on course 330 and slowed to 16 knots. Our radar picket and Tomcat destroyers were deployed in pairs, the *Brush* and *Maddox* being stationed forty miles to the northwest, and the *Colahan* and *Cassin Young* forty-five miles due west. In addition to providing us with early radar warning of approaching raids, these destroyers were to sight each of our own returning strikes visually and make sure that enemy planes were not following them back to the carriers. All returning strikes were ordered to circle the destroyers at fairly low altitude for this purpose.

At 1148* we detected a small group of bogeys bearing 280 degrees at a range of 64 miles. This position placed them approximately ten miles *ahead* of and five miles north of our own strike planes returning over the sea from Formosa. We immediately reported the bogeys to the task group fighter director in the *Essex*, fully expecting that Dutch would acknowledge promptly and order us to intercept with the *Essex*'s fighters then circling overhead on combat air patrol. The acknowledgment came, but from a voice that was clearly not Dutch's, and with no order to intercept. Almost at once, the *Washington*'s combat information center came on the air to say to the flagship, "We have that target friendly." This transmission meant that they thought the planes on that range and bearing were showing Identification Friend or Foe (IFF), and were therefore our own planes. Someone in the *Essex*—still not Dutch—acknowledged this transmission from the *Washington* on behalf of the task group fighter director, but the *Essex* apparently was waiting for her own radar operators to find the target and render their opinion as to whether it was bogey or friendly. Immediately after calling the flagship about the bogeys, I called the officer of the deck on my squawk box and alerted him. My good friend Lieutenant Stuart Carr had the deck at the time, and he in turn alerted the gunnery officer and the captain, the latter being in his sea cabin on a lower deck in the island. I now had to report to Stuart that the *Washington* said our bogeys were friendly, but we did not think so.

Three things bothered me badly. First, I knew from previous instances in the South China Sea and elsewhere that the *Essex* radar had blind spots and often could not "see" targets we could; second, new ships joining the task force were apt to be unable to distinguish between friend and foe, if the two were in close proximity; finally, having worked so long and so closely with Dutch, I knew he shared my belief in the absolute necessity of getting the fighters started immediately to intercept closing bogeys. He and I agreed completely on the importance of even a minute or two; in a case like this, the fighters could always be recalled if the bogeys proved to be friendly, but we could never regain the loss of time if they were kept on station overhead until the bogeys proved to be foes.

We were taking a new plot on the raid every minute, and both the

---

*The precise times in this account are taken from the radar plots appended to the USS *Langley*'s Action Report of 30 Dec 1944-25 Jan 1945 made available through the courtesy of the US Naval War College.

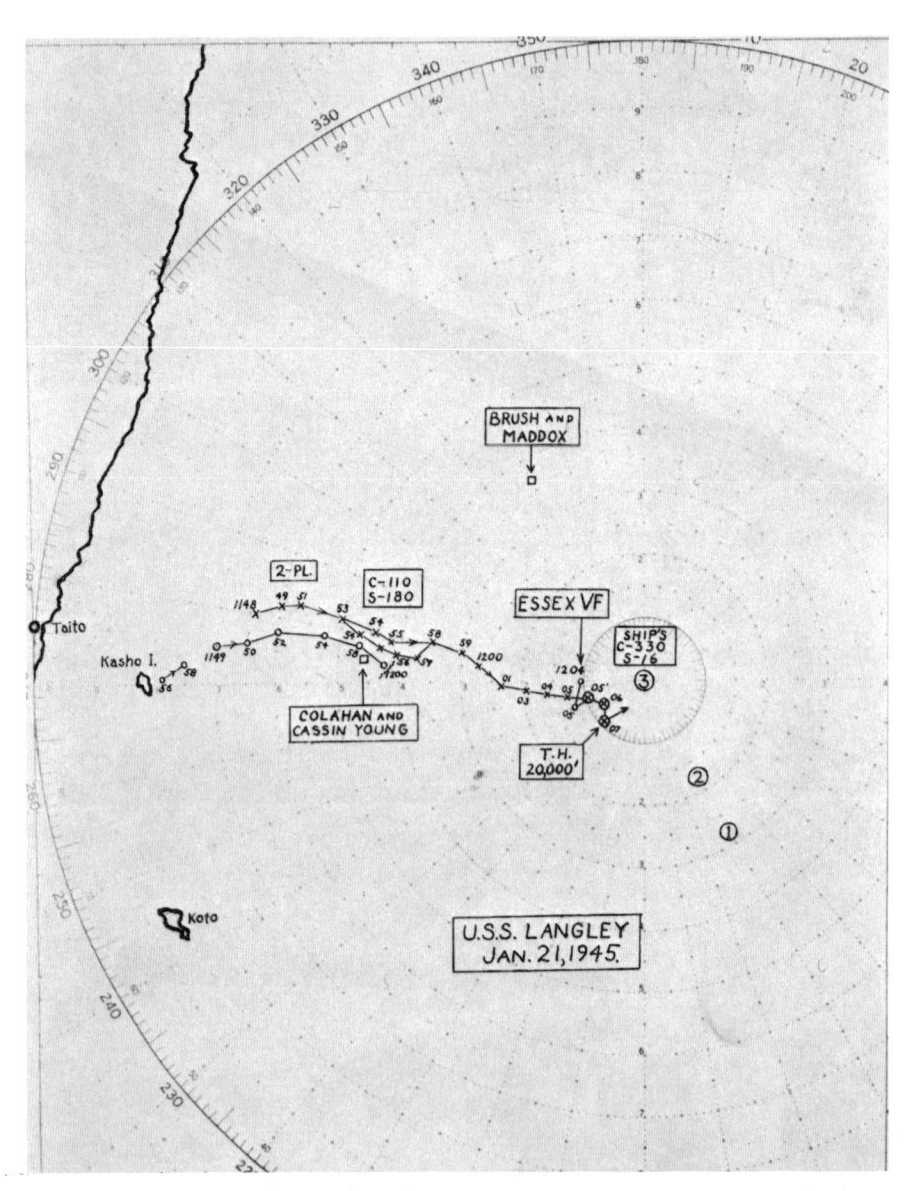

Radar plot of the suicide raid hitting the *Langley* and *Ticonderoga* off Formosa, 21 January 1945. Plot from the *Langley*'s action report.

radar operator and I double-checked to confirm that these bogeys were not showing IFF. By 1151, they had closed to fifty-five miles, and we had already lost two minutes of precious intercept time. Suspecting that the *Washington* had confused the bogeys with our own returning strike planes a few miles away, I urgently reported the new bogey position to the flagship. I also noted the position of the friendlies, which were still a few miles astern and to the south of the bogeys, but on a parallel course and only twenty-five miles from the *Colahan* and *Cassin Young*, around which they would be making a turn before proceeding on to us. This transmission brought forth another from the *Washington*, this one to the effect that they showed *both* targets friendly. Whoever was the spokesman for Dutch still did not order an interception.

Of necessity, I reported this situation to Stuart on our bridge. By then we had four plots on the bogeys, knew there were several planes, and I could report to Stuart that they were now closing us on course 110, speed 180 knots. Stuart then asked me if I thought he should call the ship to general quarters, sending all hands to battle stations and buttoning-up all watertight doors and hatches. I replied with a strong affirmative. Stuart again called the captain down the voice tube to his sea cabin, explained the situation, and asked permission to sound general quarters. After hearing the *Washington*'s latest evaluation, the captain withheld permission to sound general quarters.

At 1153, five whole minutes after we had first reported the bogeys, their blip split in two on the radar screen, one half remaining on course, the other separating and swinging two miles to the south, very close to the picket destroyers but still a few miles ahead of our strike planes. In as urgent a voice as I could muster, I gave the flagship plots on both bogey groups and another plot on our strike, the latter then descending toward the destroyers for recognition. During the next three minutes, the bogeys that had split off to the south evidently decided not to approach closer to the destroyers, turned back north, and rejoined their comrades. By 1200, our strike planes had gone down so low to be recognized that they faded from the radar, and the rejoined raiding planes were a very distinct target only twenty-eight miles away. In the *Essex*, Dutch must have been absent from the combat information center, but by now he had returned and, on sizing up the situation, immediately ordered the *Essex* fighter director to intercept the raid with the *Essex* planes on combat air patrol. Just as they were vectored out to intercept, the raid swung slightly to the south, apparently to come in from an up-sun position. Stuart again alerted the gunnery officer, and we supplied frequent ranges and bearings.

At 1205, we got our first "merged plot" indicating that our fighters and the raiding planes were at the identical range and bearing, but not necessarily at the same altitude. Unfortunately our fighters were at 10,000 feet altitude, while the raid was high above them at 20,000 feet and ready to dive.

At 1206, our gunners spotted a Zero diving in from astern and up-sun. They commenced firing. At the same time, the combat air patrol found the raid and did what it could to break it up. When he heard the first gun go off, Stuart sounded general quarters on his own initiative.

As the fighters engaged the raiders, believed to be four kamikazes and three escorts from Tainan,* the Zero diving on us released two bombs from 500 feet, and apparently tried to escape but crashed into the sea close aboard. It was not clear whether he intended to crash our flight deck and missed, or whether his plane had been hit and, at the last minute, would not respond to its controls. At any rate, one of his bombs fell into the water just off our port beam and the other struck our flight deck just forward of the forward elevator, blasting a hole through the deck fourteen feet long and ten feet wide. Fires broke out immediately on the gallery deck and the 02 deck beneath and, with men still running to battle stations all over the ship and the guns firing at a second plane coming down on the *Ticonderoga* from the same bearing, the situation was chaotic.

We were hit at 1208. I was told later that not until several more minutes had passed did the captain emerge from his sea cabin and climb the ladder toward the bridge. I was later told also that he stopped at the island level where the quartermaster was steering the ship and asked him what the situation was; and that the quartermaster replied the ship had been hit and that there were still planes in the area making strafing runs. I don't know when the captain made it from there to the bridge, but in the meantime it was left to a very able reserve lieutenant to conn the ship in a gun-firing, fire-fighting situation.

I had my hands full as well. Just after the bomb hit, not more than thirty yards from one of the uprights in radar plot, I had my nose glued to the PPI scope in front of me searching for the other planes in the raid. Looking up from the scope at one point, I was appalled to see nearly

---

*Morison, p. 180.

Lieutenant Filo Turner and fire-fighting party on the *Langley*'s flight deck after she was hit by a kamikaze's bomb, 21 January 1945. The *Ticonderoga* burns in the background. Photo from the National Archives.

everyone else in radar plot down flat on the deck in a jumble of helmets and arms and legs. It is an instinctive reaction to hit the deck when confronted with an explosion, but it angered me to see our crew down there when we still had work to do. With a roar I got them up into position, and none too soon. At 1210, two minutes after we were hit, the second Zero, with its 550-pound bomb fused to explode deep inside the ship, plunged into the flight deck of the *Ticonderoga*. The carrier's planes, armed and gassed and spotted on the flight deck for takeoff on the next strike, soon began to burn and explode.

While Sherman started to deal with this new situation, a new raid of thirteen planes came up from Luzon to attack Admiral Radford's Task Group 38.1, then about twenty-five miles southeast of us. He was able to intercept it with his own CAP, and no ships were hit.

At 1216, while we were fighting our fires and the *Ticonderoga* was fighting her much larger and more spectacular ones, we detected still another raid coming out from Formosa. This group was on bearing 300 at a range of seventy-five miles. When I reported this one to the flagship, Dutch immediately acknowledged and ordered us to intercept with one four-plane division of the *Ticonderoga*'s fighters overhead at 20,000 feet and ordered the *San Jacinto* to intercept with another division at 8,000 feet. This raid was a particularly difficult one to intercept because it faded from the screen for a full seven minutes, then seemed to be in two parts when it reappeared. The fighters controlled by the *San Jacinto* sighted it first, when it was only fifteen miles from our formation. They identified it as four Oscars and promptly shot down three of them. The fourth, however, slammed into the burning *Ticonderoga*, hit her island, and added more fuel to the fire.

While all this was going on, we in the *Langley*'s CIC were much too busy to find out just how badly our own ship had been damaged. Eventually, it developed that the bomb that hit us was a small one, perhaps 63 kilos. It was so well placed near the bomb-stowage locker and the aviation gas storage tanks that, had it been a big one, we might well have suffered the fate of the *Princeton* three months earlier. As it was, three of our crew were killed and eleven seriously wounded, and we suffered considerable damage to the beams supporting the flight deck, to water lines, and to electric circuits. We also lost several staterooms, including the one I occupied when I first came on board, and we had that huge hole in the flight deck through which the fire-fighters were playing hoses to the decks beneath.

Not long after the *Ticonderoga* took her second hit and was giving off clouds of black smoke, I had a radio conversation with Dutch on the fighter director circuit—a conversation that would have been funny in any other circumstances. Under the prevailing doctrine of the task group, each carrier's fighter director was supposed to report to the task group fighter director the numbers and call signs of the fighters his ship was going to launch and land during the next period of flight operations. The flight plan of the day called for an early launching and landing session, and Dutch ordered each of our four carriers to report in rotation. When it came our turn, I reported how many fighters were planning to land, gave him their call signs, and ended my transmission. I thought it would be obvious to him that, once we got our fires under control, we would be able to land planes on the after end of the flight deck, but certainly could not launch planes from the forward end because of the hole. It soon became obvious that it was *not* obvious. After a considerable pause, Dutch came back on the air to ask, "What are you going to launch?" It then dawned on me that he did not know we had been hit! "We are not planning to launch anything," I replied. "Why not?" asked Dutch. I was in a real quandary.

Being only one hundred miles off Formosa, there was no way I could put all our troubles on the air waves for the Japanese to hear. Yet Dutch had to know that he couldn't count on us to launch anything until we had put out the fires and patched the hole, assuming it was patchable. I got out of it finally by taking advantage of the fact that each carrier was always referred to as a "base," just as if it were on land. (At various times our code name was Rinso Base, Cosy Base, and Bronco Base.) Calling Dutch back, the best I could think of was to say, "We have a hole in the runway at this base in front of the fence. Until it's fixed we won't be launching. Suggest you get a lookout to amplify!" I knew that Dutch, an old dive-bomber pilot, would translate the "runway" into "flight deck" and the "fence" into the barrier on the flight deck, and thus realize our problem. Sure enough, a few moments later he came back to register his understanding, and I could read between the lines his great concern for his old shipmates in the *Langley*.

Our fires were soon under control and we were able to land our planes at the next opportunity. The *Ticonderoga* was not so lucky. She was a shambles, with 143 killed, 202 wounded, and 36 planes lost. Her wounded included her captain, Dixie Kiefer, who was severely burned. The destroyer *Maddox* out on Tomcat station, was also hit during the afternoon and suffered worse casualties than we did. All told, we were one sad task group by the end of the day when Admiral Sherman sent an escort to take the *Ticonderoga* and *Maddox* back to Ulithi. The *Langley*, however, was able to stay with the group, which steamed north during the night to attack Okinawa the next day.

The lessons of this miserable day, one year and one day after we had left Pearl Harbor, seemed painfully obvious to us in CIC. No admiral should embark himself and his staff in a ship that had poor radar. No watch officer under a group fighter director should hesitate to dispatch a combat air patrol because of a difference of opinion between ships as to the true nature of a bogey. And no ship in a task group should lightly contradict another's reporting of a bogey unless she was thoroughly experienced and capable of making the judgment. In battle, a captain belongs at his battle station, not in his sea cabin.

In the hushed atmosphere of our wardroom, now in use as a dressing station for the wounded stretched out on the dining tables, I reflected on how all the tragedy could have been avoided by prompt dispatch of the CAP, and promised myself that I would pull no punches when I wrote the CIC action report. We couldn't blame our angel for this debacle: it was strictly a case of failure in intership CIC procedure. I wondered bitterly how our counterparts in CIC in the *Washington* felt about their "friendlies" now.

As soon as our fires were out, the damage-control parties went to work shoring up the flight deck and putting heavy steel plates over the hole. They worked far into the night, and by morning we were able to launch our share of the morning strikes on Okinawa. Later in the day, I had good reason to wish that this had not been so. One of the main reasons for this foray to Okinawa was to obtain good photographs for the eventual assault on the island. Some forty-seven sorties were launched on photographic missions, one of which took the life of Eddie Seiler.

Eddie was not scheduled to go on the last flight of the day, and was just killing time in the ready room with friends. Up on the flight deck, one fighter plane equipped with cameras developed a problem, and rather than transfer its pilot to another plane, the ready room was asked to supply another pilot. When this message was received, Eddie said, "I'll go," and he did. Eddie's mission was to photograph Ie Shima, the little islet close by Okinawa where the famous correspondent Ernie Pyle was later killed. With the photos safely taken, Eddie decided to use up his machine-gun ammunition before returning to the ship, and dove down low to make a strafing run. An antiaircraft shell exploded in his cockpit and he never had a chance. When the flight came back to land, I noticed that Eddie was not with the other planes. After calling him several times and getting no response, I thought perhaps he was still too far away to hear me, and asked another plane high overhead to relay the call. When this relay brought no response, I was just trying again myself when one of his squadron mates came to the door of CIC and told me what had happened.

My deep sadness at the loss of this fine man and old friend was heightened by the irony of this turning out to be Air Group 44's very last combat mission.

As soon as all our other planes were aboard, we were ordered back to Ulithi, and when we arrived on 25 January, Air Group 44 was relieved by Air Group 23.

XIX
Busy Interlude in Ulithi

Upon entering Ulithi Lagoon, instead of taking our normal place in "carriers' row," we were ordered to come alongside one of the several repair ships that were anchored in a different part of the lagoon. These vessels were huge floating workshops with skilled mechanics, carpenters, electricians, welders, and metalworkers by the score. Through their herculean efforts, in which they took great pride, many a ship that had suffered storm or battle damage was quickly put back into operable condition; and many another, more seriously damaged, was at least made seaworthy for the long voyage to a navy yard at home. We located the ship to which we had been ordered, and as we made a slow approach through rows of anchored Navy cargo ships and transports, rumors spread like wildfire. Some thought we would get only emergency repairs here and then be sent home. Others thought our boilers were in such bad shape after a year and a half of steaming that we couldn't return to fight, even if the battle damage were repaired here. Still others pointed out that some of the damage done by the typhoon had not been fully corrected.

Shortly after an inspection party from the repair ship had finished looking us over, we were boarded by a working party, the like of which we had not seen since leaving Philadelphia. Armed with welding torches, rivet guns, steel beams, and steel plates, men swarmed all over the damaged areas of the ship and attacked them with a continuous uproar day and night. Before long, the scuttlebutt turned to predictions that we could, after all, be made fit to fight.

To add to all the confusion, Air Group 44 packed up and said their farewells, and Air Group 23, led by Commander Don White, came aboard. There followed the usual meetings to get acquainted with a new group of fighter pilots and to familiarize them with the individual procedures of the task group and the ship. Other important changes were going on simultaneously. At the top level of the fleet and the task force, Halsey and McCain were relieved by Spruance and Mitscher. I remember having very mixed emotions about the top command. When Halsey commanded the fleet, I always had the feeling that, though he might get us all killed in the process, he would run risks and take chances to get the war over with as fast as humanly possible. I felt that Spruance was much too cautious, and, in Navy parlance, "kept his finger on his number," meaning that, regardless of opportunity, he would never alter a set plan if he might be criticized for doing so. I therefore much preferred to sail under Halsey. History has since mellowed my opinion: if Spruance was too cautious, no doubt Halsey was too impetuous and apt to go off half-cocked. Between the task force commanders, however, I never doubted that Mitscher was far superior to McCain, perhaps because of my previous acquaintance with the latter when he took over the Bureau of Aeronautics. Strikingly different as all these men were from each other, we were probably lucky to have them all.

At the time of the changeover of command on 26 January, Halsey sent an emotional and heartfelt message to all his ships. Less formal and more characteristic was the one I heard him send over the TBS, in which he concluded with, "The drivers may change, but the horses go on!"

Several important changes were made in the *Langley* crew, as well. Tom Smith, by then my roommate, had made lieutenant commander, was relieved as air plot officer by Loren Hickerson, and was designated communications officer of the ship. A few radarmen were available to relieve some of our hard-working group, and instead of transferring our least effective men, I did the opposite and released a couple of our best. I felt they had earned it, and the effect on all remaining members of the radar crew was salutary, to put it mildly!

The most important change in the *Langley*, however, was that her executive officer, Commander Ned Hannegan, was finally relieved by a newcomer, Commander W.C. Wingard. Hannegan had been with us from the beginning, first as air officer, then as exec. He was a fine, fair, dedicated officer, and had the respect of every one of us in the wardroom. Normally I would have hated to see him go. As it was, I was delighted for his sake because I thought that he and the air officer, Bill Guthrie, were being pushed by the captain beyond reasonable limits of endurance. Under his extreme criticism of even small mistakes by members of the crew, there was no way for the exec or the air officer to escape.

Once, when I was the Air Department duty officer, the captain had Hannegan awakened at three in the morning and ordered him to get me off the flight deck, where I was supervising the loading of a new torpedo bomber from a lighter alongside, and bring me to his quarters immediately. En route, Ned said to me: "I don't know what the trouble is, but whatever it is, just stand at attention and take it. Don't argue with him, and don't offer any excuses. It'll only make it worse." I followed these instructions to the letter while, for about twenty minutes Wegforth gave me the most intense dressing-down I ever had before or since. The reason? He had awakened when the lighter came alongside, and, as we hoisted the TBM up over the side with our flight-deck crane, he had spotted a member of the lighter's crew smoking a cigarette. This enraged him and, to hear him tell it, I had jeopardized the entire ship and all its crew by not having seen the smoker and ordered him to stop. All this despite the fact that the airplane was unarmed and de-gassed, and the perpetrator of the crime was on *his* vessel, not *ours*! When he was finally through with me, the captain went on to make clear to Ned what he thought of allowing such a nincompoop to stand Air Department duty watches.

He did not, however, order me removed from the watch list and I did not harbor any grudge against him for the lecture. Under his style of command, I well understood that if you were caught in an error, you caught hell for it. I was, however, both sorry and embarrassed that I had added to Commander Hannegan's burdens, and apologized to him for it.

As the criticism rolled over Hannegan and Guthrie day after day, they suffered in silence. Both became gaunt. Commander Hannegan began to lose his hair and his eyebrows, first in patches, then in toto. Commander Guthrie became almost a recluse, heading for his stateroom when flight quarters were over, and having his meals served there instead of in the wardroom. They were the most conspicuous targets of us all, and the most severely criticized. We gave Ned Hannegan a rousing sendoff when he at last made his escape to a new assignment and eventually to promotion to captain and rear admiral.

Our problems with the captain were not lost on the medical officers in the ship. After witnessing some of his outbursts, hearing about others, and seeing their effect on some of the officers, they became concerned both for the captain and for those serving under him. My private opinion, kept strictly to myself, was that he was in serious need of psychiatric help; and I kept hoping the medical officers would get him relieved. Forty years later, I was told by one of the senior medical officers that he and a colleague *had* tried to get the captain relieved by higher authority; but were unsuccessful, apparently because the operating record of the ship was so good.

On a happier note, I had a very pleasant experience while we were alongside the repair ship. One day while I was taking a nap, I awoke with the feeling that someone was looking at me, and sure enough a stranger's head was peering through the flameproof curtains at my door.

"Are you John Monsarrat?" said the head.

"Yes," said I.

"Well, I'm Tom Wheeler, and I'm married to Natalie Howard. Nat said to be sure to look you up when I came out here. I'm the navigator in one of the transports anchored near here, and saw the *Langley* coming in the other day. So here I am, and how are you?"

Natalie was a good friend from Columbus and this was like getting a letter from home. Tom had brought me a present of a bottle of Scotch, and we had a long and pleasant visit before he had to take his boat back to his ship. Later on, he came back to see a movie on our hangar deck, and I welcomed the chance to get away from the racket in our ship and have dinner in his. It was an amazing coincidence that he found me. I did not see him again until, many years later, we met at his stepson's wedding in Boston, at which time I welcomed the chance to repay him with another bottle of Scotch.

In about ten days, the workmen were through with us, and we were declared fit to fight again. We cast off and steamed over to take our place in carriers' row. Issues of cold-weather clothing began coming aboard, and soon the new operations plan confirmed that we were at last going to Japan!

XX
Raids on Tokyo in Support of Iwo Jima

On 26 January 1945, TF 38 reverted to TF 58. For the assault on Iwo Jima, Task Force 58 was restructured, this time into four daytime groups and one night fighter group. The latter was composed of the *Saratoga*, which had returned from stateside repairs to battle damage, and the *Enterprise*. The *Langley* was moved from Admiral Sherman's group to Admiral Radford's Task Group 58.4: the other carriers in our group were the *Yorktown*, the newly arrived *Randolph*, and the *Cabot*. Our main assignment during the operation was to hit Tokyo and its nearby airfields, in order to prevent the Japanese from sending raids from Honshu down to Iwo Jima to attack our landing forces. While we were to strike Iwo Jima and Chichi Jima as part of the prelanding bombardment, this time the provision of air support to the Marines, once landed, was to be left mainly to the escort carriers accompanying the transports.

By this time, a directive had come from CinCPac redefining the scope of the combat information center in carriers. Heretofore, CIC had really been synonymous with radar plot. Henceforward, according to the directive, CIC was to include radar plot, air plot, air combat intelligence, aerology, and the photographic laboratory. It was also stipulated that the executive officer should shift his battle station to radar plot, where he would be able to get the best possible idea of the situation if the ship were hit and the captain disabled on the bridge. In our case, this presented no particular problems. Commander Hannegan had made radar plot his battle station for the last several operations, and Commander Wingard now did the same. We had long worked in harmony with the other elements in the ship, and simply continued to do so pretty much as before. Right after receipt of the directive, however, the captain had the officer of the deck call me, instead of the aerologist, to ask, "Why is it raining?"

Raids on Tokyo 145

```
AIR DEPARTMENT OPERATION PLAN FOR FRIDAY 16 FEBRUARY 1945
                                              SUNRISE 0719
    D-3 DAY      TOKYO                        SUNSET  1818
    Air Dept. Duty Officers - LT. J. ALLING until 1200 Feb. 16th.
                              LT. J. MONSARRAT   "     "   "  17th.
    TASK GROUP AIR PLAN
       0635 - RANDOLPH launch 4 VF RAPCAP #1.
              LANGLEY launch 8 VF CAP #1.
              CABOT launch 12 VF CAP #1.

       0645 - YORKTOWN launch 16 VF Sweep #1A; 4 VF T.C.#1.
              RANDOLPH launch 16 VF Sweep #1B; 4 VF RCAP #1.

       0715 - YORKTOWN launch 16 VF Sweep #2A.
              RANDOLPH launch 12 VF Sweep #2B.

       0830 - YORKTOWN launch 16 VF Sweep #3A.
              RANDOLPH launch 12 VF Sweep #3B; 4 VF T.C.#2; 4 VF RCAP#2.

       0930 - YORKTOWN launch 12 VF Sweep #4A; 4 VF RAPCAP #2.
              RANDOLPH launch 12 VF Sweep #4B.
              LANGLEY launch 12 VF CAP #2.
              CABOT launch 8 VF CAP #2.

       0945 - YORKTOWN land 16 VF Sweep #1A.
              RANDOLPH land 16 VF Sweep #1B; 4 VF RAPCAP #1.
              LANGLEY land 8 VF CAP #1.
              CABOT land 12 VF CAP #1.

       1045 - YORKTOWN launch 16 VF Sweep #5A.
              RANDOLPH launch 16 VF Sweep #5B.

       1100 - YORKTOWN land 16 VF Sweep #2A; 4 VF T.C.#1.
              RANDOLPH land 12 VF Sweep #2B; 4 VF RCAP #1.

       1130 - YORKTOWN land 16 VF Sweep #3A.
              RANDOLPH land 12 VF Sweep #3B.

       1230 - YORKTOWN launch 16 VF Sweep #6A; 4 VF T.C.#3; 4 VF
              RAPCAP #3.
              RANDOLPH launch 16 VF Sweep #6B; 4 VF RCAP #3.
              LANGLEY launch 8 VF CAP #3.
              CABOT launch 12 VF CAP #3.

       1245 - YORKTOWN land 12 VF Sweep #4A.
              RANDOLPH land 12 VF Sweep #4B.
              LANGLEY land 12 VF CAP #2.
              CABOT land 8 VF CAP #2.

       1345 - YORKTOWN land 16 VF Sweep #5A; 4 VF RAPCAP #2.
              RANDOLPH land 16 VF Sweep #5B; 4 VF RCAP #2; 4 VF T.C.#2.

       1430 - YORKTOWN launch 20 VF Sweep #7A; 4 VF RAPCAP #4.
              RANDOLPH launch 20 VF Sweep #7B.

       1530 - YORKTOWN land 16 VF Sweep #6A; 4 VF RAPCAP #3.
              RANDOLPH land 16 VF Sweep #6B.
              LANGLEY launch 12 VF CAP #4.
              LANGLEY land 8 VF CAP #3.
              CABOT launch 8 VF CAP #4.
              CABOT land 12 VF CAP #3.

       1745 - YORKTOWN land 20 VF Sweep #7A; 4 VF T.C.#3; 4 VF RAPCAP#4.
              RANDOLPH land 20 VF Sweep #7B; 4 VF RCAP #3.
              LANGLEY land 12 VF CAP #4.
              CABOT land 8 VF CAP #4.

              YORKTOWN has VFN Duty.
              RANDOLPH has Night Duty Deck.
```

Air operation plan for the first strikes on Tokyo by Navy planes in support of Marine landings on Iwo Jima.

No Navy planes had yet struck Tokyo, although the Army Air Forces had begun to hit it with B-29s from Guam and Saipan. Task Force 58 sortied from Ulithi on 10 February and made its way north under the strictest radio silence. Bad weather prevented enemy scouting planes from taking to the air and helped keep our movements secret: the picket boats we encountered were very quickly dealt with. By dawn on 16 February, we had arrived unscathed at our launch point, only sixty miles from shore.

On the first day of the strikes, while I was the Air Department duty officer for the *Langley*, our mission was to provide combat air patrol over the task group, while other carriers hit the airfields around Tokyo and searched for shipping targets. The enemy sent large numbers of fighters to intercept our strike planes but, probably because of the very bad weather, did not mount retaliatory strikes against the carriers so very close to the shore of Honshu. While the strike planes had a hard day in the bad weather, with heavy air opposition and heavier antiaircraft, we had it surprisingly easy.

Not so for our air group next day. It was their turn to help supply the strike force that was flown against enemy aircraft and aircraft engine plants. Over Yokohama, Don White was shot down by a Zero on his first strike as CAG-23 and we lost a fine leader. His wingman saw him bail out, and he was recovered from prison after the war.

With the weather so bad, it was hard to estimate damage, but a few ships were sunk in Tokyo Bay and the task force claimed 341 planes in the air and 190 on the ground. The cost to it, however, was very high; 60 planes and many pilots. On the way back to Iwo Jima, we attacked the airfield at Chichi Jima, sank several small ships, cratered the runways, and then half the force refueled while the other half took station off the island to support the landings on the nineteenth.

For three days after the landings, we augmented the air support the CVEs were providing the troops, and each night retired a little farther from the island to ward off enemy air attacks. When they came, they came with great pyrotechnics. First, the "lamplighters" dropped bright magnesium flares on the western side of the carrier force. Then, for the first time in our experience, the Japanese made good use of "window," small strips of metal foil cut to match the wave lengths of our radar and dropped in bundles just to the east of our ships. When the strips fanned out in the air and slowly floated down, their radar echos were so strong that they blocked out our radar vision in that particular sector. Thus, with our ships brilliantly silhouetted against the flares and float lights to the west, and our radars blind to the east, the enemy torpedo planes had a golden opportunity to hit us with devastating effect. For some reason, they did not take it. Time after time, I watched on the radar as groups of torpedo planes came straight at our blind spot from the east, penetrated it, then stopped before they got all the way through it to reach us. If they had known how blind we were, they would have been a great deal bolder.

Closer to the island, they did press home their attacks on our ships. We had lent the *Saratoga* to cover those forces with her night fighters. One night, she took no less than five bomb and torpedo hits. She survived, but limped away, never to return. (When the US flag first went up on Mt. Suribachi, I heard a patrolling fighter make the announcement, little realizing that this event would become the subject of the famous memorial sculpture by Felix de Weldon in Washington.)

On the twenty-second, we left to return to Tokyo, but the weather was so foul that our strikes had to be canceled. After taking more photographs at Okinawa, the task force returned to Ulithi on 4 March.

XXI
Fast Carriers Versus Kamikazes at Okinawa

The plan for Operation Iceberg was delivered to the ships of Task Force 58 while they were at anchor in the lagoon at Ulithi, replenishing and regrouping after the capture of Iwo Jima. It was the second week in March 1945, and the task force was at peak strength: seventeen fast carriers, eight battleships, eighteen cruisers, and sixty-three destroyers.

Few things are as demonstrative of high professionalism as is a Navy operation plan. The planners' ability to foresee and provide what will be needed, and where and when, in ships, in armament, in logistical support, and in communications, involves visualizing and coordinating thousands of details. That the Navy's planners did all this was to me one of the miracles of the Pacific war. I wish that one of the Pacific operation plans might sometime be published so that the civilian world could get a better appreciation of what naval planners contribute to our nation's security.

It came as no surprise that the capture of Okinawa was to be our next objective; nor did we have any illusions about its difficulty. The role of the fast carriers, as outlined in the op plan, was the familiar one we had already played in fifteen months of consecutive assaults on the Marshalls, New Guinea, the Marianas, the Palaus, the Philippines, and Iwo Jima. Reduced to simplest terms it was to get to the target area well ahead of the invasion fleet, destroy as many as possible of the enemy aircraft within range, strike hard at the target itself, provide close support for the initial landings, and be prepared to intercept the enemy's reaction forces in the air and at sea.

From our experience with the kamikazes ever since their first appearances off the Philippines, it was obvious that they would be out in full force once we landed at Okinawa, on the doorstep of the home islands. The main question was how many planes and trained pilots the Japanese would have left with which to oppose us. Estimates ran between 2,000 and 3,000 from the more than 120 airfields within range of Okinawa.*

At this stage of the war, Ulithi was considered a safe anchorage for the Fifth Fleet in the brief intervals between major operations. The Marines at the air station ashore, equipped with their own radar, fighters, and reconnaissance planes, assumed responsibility for protecting the atoll and the anchorage from air attack. This arrangement allowed the fast carriers and their escorts to secure their own radars when they entered port for a few days of much-needed maintenance and repairs. Ships at anchor were allowed to show lights at night, and even to show movies on deck after dark.

On the evening of 11 March, the third day before we were scheduled to leave Ulithi, the *Langley* and her neighbor, the carrier *Randolph*, had movie screens rigged. Ours was on the flight deck and I arrived early with Tom Sorber to get a good seat. As we chatted before the movie began, Tom pointed up to the sky above our bridge and said to me, "Look at that idiot flying around up there without any wing lights." Sure enough there was an airplane circling overhead and in a matter of moments, without any warning whatsoever, "that idiot," a fully armed kamikaze, dove his airplane and his bomb into the *Randolph*. After a huge explosion, there was a stunned silence for a moment and then pandemonium as ship after ship went to general quarters, darkened ship, got up steam, and manned the guns.

Tom and I dove into CIC where we were keeping a communications watch even though our radars were secured. After many false alarms and wild reports things began to settle down, and ships eventually went back to normal routine, albeit carefully darkened down. Not until long after the war did we learn that the kamikaze was one of twenty-four Frances bombers assigned to the Azusa Special Attack Unit which had been organized in the home islands specifically to attack our fleet in Ulithi.** Thirteen developed engine trouble and dropped out on the 1600-mile flight down from Kanoya, on Kyushu, and, of the remaining eleven, the one that hit the *Randolph* was the only one to score. One was found to have crashed into one of the small islets surrounding the

---

*Morison, v. XIV, p. 89.
**Rikihei Inoguchi et al., *The Divine Wind* (Annapolis: US Naval Institute, 1958), pp. 132-3.

lagoon, apparently having mistaken it in the dark for a carrier flight deck. No explanation ever reached us as to why the Marines ashore had failed to detect the raid, and if the other nine bombers actually made it to Ulithi; why they missed so many juicy, *anchored* and *lighted* targets is a mystery. Twenty-seven members of the *Randolph*'s crew were killed, and both her flight deck and her hangar deck were badly holed.

Task Force 58 sortied from Ulithi for Operation Iceberg on 14 March. The ships were organized into four task groups operating in concert within about twelve miles of one another under the command of Vice Admiral Mitscher in the *Bunker Hill*. He in turn reported to Admiral Spruance as Commander, Fifth Fleet, in the *Indianapolis*. The *Langley* remained in TG 58.4 commanded by Rear Admiral Radford in the *Yorktown*. In accordance with the plan, TF 58's first assignment was to attack the airfields on Kyushu, Shikoku, and the environs of Tokyo, in order to weaken the Japanese air forces as much as possible before the landings on Okinawa on 1 April.

During the early hours of 18 March we approached to less than 100 miles from the coast of Kyushu, and at dawn began launching successive strikes, ranging farther and farther inland. In retaliation, the Japanese attacked the task force all day long with conventional bombers as well as kamikazes. The brunt of these attacks from dawn to dark fell upon TG 58.4.

On the first day of Kyushu strikes, Raid 1 appeared on the board very early in the morning, and by nightfall we were up to Raid 51, the highest number we ever faced in a single day. The raids followed no particular pattern. Some were single planes such as the Judy that our CAP intercepted 30 miles away. Others came in groups of eight or ten. Some were conventional dive-bombers escorted by fighters; others were kamikazes. The tactics used by the attackers varied all the way from wave-top approaches, so as to sneak in beneath the radar, to approaches from above 15,000 feet.

Launching and recovering our own strikes on Kyushu and vectoring the CAP out to intercept raid after raid, kept the carriers in TG 58.4 working all day at a furious pace. Our attackers had obviously been ordered to "get the carriers," and there was nothing new in that. But the sheer numbers of the raids and the diversity of approaches raised the tension in CIC even higher than usual. As the track of a new raid started to close in on the center of the plot where we were located and as the track of our CAP crept out to meet it, the atmosphere in CIC became electric.

Particularly, as far as the kamikazes were concerned, there was a tremendous premium in intercepting far enough out from the ship to give the fighters time to destroy or at least disperse the raid. As noted elsewhere, a raid traveling at 300 miles per hour would cover the last 50 miles to its target in ten minutes. Thus, if defending fighters intercepted it a full 50 miles away, they would have ten minutes in which to deal with it. However, in order to intercept a raid 50 miles away, the defensive force would have to start its interception run while the raid was 100 miles away, assuming that both it and the enemy were traveling at 300 miles per hour. Except in extreme cases, our radar could not detect an incoming raid that far away, and, the lower the altitude of the raid, the closer it could approach before we got our first indication of it. Consequently, most raids were intercepted closer in, there was less time for the fighters to work them over, and the fighter directors bore the grave responsibility of making split-second decisions and making them accurately to save a precious moment or two.

Our record for that long day was far from perfect. We were confident that we had intercepted and destroyed the majority of the attackers. But too many survived to fly home, and too many penetrated our fighters and our screen of antiaircraft fire. None actually crashed into one of our ships, but the *Yorktown* was hit by a bomb from a Judy, suffering 26 casualties; and the *Intrepid* had fires on her hangar deck from burning pieces of a Betty shot down while attempting a suicide run on her.

On the more positive side, the task force's search planes found impressive shipping targets, including the huge battleship, *Yamato*, along the coast of Honshu, and its strikes scored heavily on the airfields of Kyushu. But all in all, 18 March was a sobering day for a veteran CIC group that conditioned itself to believe that it had failed every time an enemy aircraft penetrated within range of the task force's guns.

The next day began even more tragically with the agony of the *Franklin* and the ordeal of the *Wasp*. A little after seven in the morning, when the *Franklin* was launching a strike, she was hit by two bombs from a plane that no one had detected. One bomb exploded on her hangar deck, setting fire to the many airplanes parked there, some of them armed; and the other exploded on her flight deck, setting fire to the fully armed planes warming up to get out on the strike. One by one, the burning planes themselves exploded, fires and smoke enveloped the entire ship, and two of the three elevators were completely wrecked.

Only two minutes later, another attacker, similarly undetected, dropped a well-placed bomb on the *Wasp* in the same task group as the *Franklin*, TG 58.2. The *Wasp* had just secured from general quarters and fortunately most of her planes were out on strike. Her damage-control parties quickly got the resulting fires under control, but 101 men were killed and 260 wounded.

The *Franklin*, on the other hand, had a frightful time getting her fires under even a semblance of control, and Rear Admiral Davison, the task group commander, advised Captain Leslie H. Gehres to prepare to abandon ship. Instead, the captain, aided by the cruiser *Santa Fe* and towed by the cruiser *Pittsburgh* began the tortured, 12,000-mile journey to New York. She was unquestionably the most heavily damaged carrier to survive, and her casualties numbered 724 killed or missing and 265 wounded. Among the former was her entire CIC team, caught in fires so hot that when they were finally extinguished the compartment was fused together.

Task Group 58.2 had been operating just to the north of us in TG 58.4, on the bearing of Japan. As we continued north, the *Franklin* came drifting back through our group and it did not look as though she could possibly survive. To give her a chance, however, we obviously had to intercept the raids that were sure to follow, as long as she was still afloat. In this we were successful throughout another long day; and our attack aircraft reported successful strikes against shoreside targets and the shipping discovered the previous day.

For the past year and a half, the fast carrier task force had operated under a doctrine of refueling at sea every third or fourth day in order to be ready to go at flank speed in pursuit of enemy ships, if and when they emerged. Usually, we refueled one task group at a time, and during Operation Iceberg the service force had been expanded so that we could also rearm and even reprovision at sea. When it was our turn, the normal procedure was for us to retire after the last strike of the day, steam away from the target all night, and rendezvous with the replenishment group at dawn the next day. One by one, the carriers would come alongside the tankers to take on oil and aviation gasoline; then they would proceed to the ammunition ship for bombs and bullets, and to the provision ship for food and other supplies. Meanwhile, the destroyers would be refueling from the battleships while they and the cruisers awaited their turns at the tankers. Finally, the escort carriers accompanying the service fleet would fly replacement planes and pilots to the fast carriers as needed. In this fashion, the fast carrier task force could and did keep operating at sea for three months without respite, although it was 4,000 miles from Pearl Harbor and 6,000 from San Francisco.

Now came the debut of the baka bomb.

Naval Intelligence had been warning us of this weapon for some time. Called oka (cherry blossom) by the Japanese, the baka was a 4,700-pound flying bomb with stubby wings and a tail, not unlike a miniature fighter plane. It was built of wood, powered by rockets to attain speeds of up to 600 miles per hour, far in excess of any fighter then in existence, and designed to accommodate a single kamikaze pilot who would fly it full speed into a targeted ship. This device was the invention of a Japanese aviator, Ensign Ohta,* who conceived it as a weapon to be slung underneath the twin-engined Betty bomber, which would carry it to within ten or twelve miles of its target. While this concept overcame the problem of the baka's extremely short range, its weight and drag severely limited the Betty's ability to maneuver and defend itself. None of these details were known to us at the time. Nor did we know that the pilot of the baka rode in the Betty until it was time for him to enter his own craft for the final dive.

Early in the morning of 21 March, Japanese search planes located the task force off Kyushu and their bombers were ordered to launch a baka attack. We in the *Langley* were the first to detect the approach of this raid coming in from the northwest and about 85 miles distant. We could tell immediately that it was a large raid and reported it as such in our first call to the group's FDO. He immediately ordered us to intercept with one division of the *Intrepid*'s fighters from the combat air patrol. Within minutes, as the raid began to appear on the radar screens of the other ships in the task force, the task group commander ordered the launching of additional CAP.

We vectored out the CAP on the bearing of the incoming raid and at "Buster, Angels eighteen," fighter director code for full sustainable speed and 18,000 feet altitude. On their way out, we gave them our estimates of the size of the raid and the altitude at which it was flying. When they came within a few miles of the enemy, just as we were preparing to order them to orbit so as not to overshoot, they called in their "tally ho," and described the raid as "many hawks and rats" (FDO code for bombers, and fighters).

Interception took place at a range of 60 miles from the carriers,

---

*Inoguchi et al., p. 140.

giving our fighters the precious time they needed to do their job. The next divisions of the CAP to arrive were fighters from another carrier, as were most of the subsequent reinforcements soon launched and vectored out as backup and support. The voices of the division leaders were therefore unfamiliar to us and their transmissions had the complications that resulted from their need to use throat microphones and oxygen masks. Nevertheless, a moment or two after their initial contact, we sensed a rising excitement on the part of the pilots that was over and above the natural adrenalin-pumping to which we were all accustomed. They were seeing baka bombs for the first time, and all hands concentrated on the Bettys carrying them. Some were shot down while they still had their load. Others saw what was coming and, in self-defense, jettisoned their baka into the sea. As reinforcements arrived to help the first groups of CAP, the excitement rose. The escorting Japanese fighters were engaged, and in a furious fifteen minutes all the Bettys and all but one or two escorting fighters were "splashed." None penetrated within thirty miles of the task force.

When the CAP returned to the decks of their carriers, their debriefing by the air combat intelligence officers and the evidence of their gun cameras gave us the first hard evidence of what we were up against in the baka. We know now that there were eighteen Bettys and thirty fighters in the raid.*

In subsequent operations at Okinawa, other bakas appeared, and at least one of the small ships supporting the invasion force was sunk by a baka; but their disastrous failure on 21 March caused great gloom among their specially trained force, and undoubtedly limited their further use.

On 22 March, with TG 58.2 escorting the crippled *Franklin* and the *Wasp* out of the danger zone, the task force was reorganized into three task groups, instead of four. The fast carriers then shifted from their attacks on Kyushu and Honshu to strikes on Okinawa proper. Shortly after midnight on the twenty-second, the destroyer *Haggard*, on her night picket station in advance of TG 58.4, obtained a sonar contact. She and the *Uhlmann*, also on picket station, both had surface radar contacts with something earlier in the night, but the *Haggard*'s sonar contact now confirmed that a submarine was in the area. She immediately dropped a depth-charge pattern. A few moments later

---

*Toshiyuki Yokoi, *The Japanese Navy in World War II* (Annapolis: US Naval Institute, 1969), p. 132.

Lieutenant Walter B. Woodson, Jr., executive officer of the *Uhlmann*, with his battle station in CIC, reported a radar contact very close alongside the *Haggard*. So close was the contact that, from the radar standpoint on board the ship, it was invisible, lost in the ship's own "sea returns." However, alerted by the *Uhlmann*'s radar as to just where to look, the *Haggard* saw the submarine close aboard, rammed it, and sank it. Having considerably damaged her bow in the ramming, the *Haggard*, with the *Uhlmann* as her escort, was detached from the screen and sent to Ulithi for repairs.

With the landings scheduled for 1 April, our objective was to destroy as many as possible of the Japanese airplanes and air support facilities on Okinawa. From first light until last, we conducted bombing and strafing operations, and the night fighters kept up some pressure after dark.

The Japanese air forces responded to these attacks with less than full fury. Knowing full well that a massive invasion was coming soon, and having already lost a considerable number of planes and facilities to our strikes on Kyushu, their high command committed only a fraction of its kamikaze strength against the fast carriers. They held back the majority for major assaults against the landing forces. The Japanese planners had prepared a massive attack plan called Ten-Go, which called for the use of up to 4,000 conventional and suicide planes from Navy and Army sources on Kyushu and Formosa; suicide boats from Okinawa and the Kerama Retto; and a last sortie of their remaining warships from the Inland Sea. According to an agreement reached in February between the Army and the Navy:

"In general Japanese air strength will be conserved until an enemy landing is actually underway or within the defense sphere. The Allied invasion force will then be destroyed, principally by Special Attack (Kamikaze) units . . . . The main target of Army aircraft will be enemy transports, and of Navy aircraft, carrier attack forces."[*]

The American plan called for the assault and occupation of Kerama Retto on 26 March, and on Easter Sunday, 1 April, a feint against the southeastern coast of Okinawa and the main landings were to be made opposite two of the airfields on the western shore. Kerama Retto is a group of small islands only a few miles to the west, and their value was to serve as a roadstead in which damaged ships could seek sanctuary,

---

[*]James H. and William M. Belote, *Typhoon of Steel* (New York: Harper & Row, 1970), p. 31.

and unloading transports and supply ships could regroup after dispersing from the landing areas. Altogether, 182,000 troops and their supplies were to be landed from 433 assault ships in the largest operation of the Pacific war.

A very important part of the American operation plan was the most extensive use yet of radar picket ships. Usually destroyers, these ships were ordered to take station at sixteen points offshore and to report to Admiral Kelly Turner's command ship, the *Eldorado*, designated Delegate Base, whenever they detected enemy aircraft approaching. Each destroyer had a fighter director team on board, and they were to direct fighters of the combat air patrol which to begin with, would come from the carriers, and later from captured airfields on the island. By giving early warning and initiating early intercepts, it was hoped that we could cut down the number of kamikazes that penetrated to the landing ships off the beaches.

Additionally, TG 58.4 and other task groups as well, supplied its own radar pickets by stationing one or two destroyers from its screen miles in advance of the task group and providing them with their own CAP, called the RAPCAP for Radar Picket CAP. At night, these destroyers returned to their stations in the screen.

On 26 March, after a heavy bombardment from the invasion fleet, the troops occupied Kerama Retto, meeting relatively light resistance. In the process, the Kerama base for suicide boats was overrun, reducing some of the opposition the main landings would have to face. The picket destroyers were stationed the same day and supplied during daylight with combat air patrol from Admiral Durgin's escort carriers.

To maintain pressure on the home bases of the kamikazes at Kanoya and other fields on Kyushu, B-29s from Saipan scheduled large raids on 27 and 31 March, and Task Force 58 left its Okinawa station to hit Kyushu again on the twenty-ninth. During these Kyushu strikes, we directed four fighters from the *Langley* that were part of the CAP to intercept an incoming kamikaze. The fighters did not get into shooting position until the suicide plane was only a few miles from the destroyer screen and closing fast. Under fighter direction doctrine, the controlling FDO was obligated to warn fighters in hot pursuit when they and their target were coming in range of ship's antiaircraft fire. The fighter director code for this warning was the simple word *guns*, which meant "break off the engagement and get back out of gun range." Whenever a new squadron came on board, this was one of the points we emphasized when we briefed them on fighter direction procedure. This time, however, one two-plane section of the CAP, with machine guns blazing, continued on the tail of the smoking kamikaze and followed him right into his final dive. By this time, the air was saturated with antiaircraft fire from virtually all ships inside the screen, and our own two planes were destroyed along with the "bandit." Among our many sad losses this was one of the hardest of all to take.

That evening after flight operations were over for the day, the executive officer convened an inquiry. I was asked to testify under oath whether or not I had warned the fighters off as they approached the screen. We had had a frantically busy day in CIC, and whether it was from exhaustion, or distress, or the number of communications I had dealt with, or perhaps a combination of all three, I could not clearly remember just what I had said to these fighters. The Communications Department then produced the log of my transmissions on that intercept. It showed that I had warned them off not once but three times. I felt greatly relieved, but my relief did nothing to diminish the sorrow we all felt at losing those brave men.

Early next morning we directed the interception of an Oscar bound for Okinawa.

Finally the stage was set for 1 April, "Love Day," the scheduled landing at Okinawa.

The assault troops hit the eight-mile landing beach precisely on time, the culmination of voyages that had started from eleven ports all across the Pacific, from the coast of California to Leyte. The landing was preceded and followed by huge salvos of 16-inch and 14-inch gunfire from the battleships, and intense strafing and bombing from the supporting carriers. At first, opposition was surprisingly light, casualties low, and retaliation from the air gratifyingly weak as a result of all of our efforts during the preceding two weeks.

This soon would change!

For the first few days after the landings, Japanese air attacks were sporadic and not very well coordinated. Then, on 6 April, the blow fell.

It fell in the form of Kikusui No. 1, the Japanese name for the first of ten massive air attacks directed from Kanoya on Kyushu. *Kikusui* means "floating chrysanthemum," and for the first one the high command had readied 699 planes, 355 of which were kamikazes, a distinction that soon became blurred by an order to all the pilots to consider themselves kamikazes, whether or not they had been specifically trained in this tactic.

Most of the attacks were mounted from the airfields on Kyushu and Formosa, and throughout the afternoon they concentrated on a wide area and variety of targets. Waffles, which in fighter director parlance meant the degree of cloud cover, was very high and favored the attackers over the defenders because it provided good cover on their approach. However, the weather prevented approximately half of the 120 attackers assigned to hit TF 58 from finding our formation, and they went on to Okinawa to add their strength to the hundreds of others assigned to strike ships there. None of the 60 or so that did find and attack TF 58 hit a carrier, despite several close calls, and most were successfully intercepted by the CAP.

Closer in to Okinawa, the situation was quite the reverse. The attackers, vastly outnumbering the fighter protection provided by the fast carriers and the escort carriers, bore down on the radar pickets controlled by Delegate Base. Despite heroic efforts, there was simply no way the pickets and their CAPs could prevent the attacks from penetrating to the many supply ships of the expeditionary force off the landing beaches and to the ships assembled in the roadstead of Kerama Retto. Admiral Turner estimated that 182 planes in 22 groups attacked the expeditionary force during the day. Of these, he believed that only 55 had been shot down by the CAP, 35 by ships' antiaircraft fire, and 24 destroyed in crash dives. Task Force 58 claimed to have shot down 249 planes, including those it caught over the island. Nevertheless, three destroyers, one LST, and two ammunition ships were sunk, and ten other American ships were damaged. Making full allowance for the facts that our shore-based aircraft facilities were not yet operational and that this was the first (and as it turned out the most severe) test of the complex radar picket system around the island, the results of the day were extremely worrisome.

Having directed fighters from a destroyer myself during experiments off Hawaii in 1943, I was especially sympathetic with the problems of the fighter directors in the pickets. In the cramped quarters of a destroyer, not much space could be allocated to a CIC. The fighter direction teams had to be small, and the radar and other equipment for fighter direction was in no way comparable to that available in the carriers. Understandably, those in control of the pickets asked immediately that many more destroyers be assigned to the duty, so that instead of one to a station, there might be several to handle the load and provide mutual antiaircraft support. But, at this juncture, it was not possible to spare ships from other pressing duties, and most of the destroyers had the lonely duty of manning their stations with only the CAP for support.

Intelligence acquired through radio intercepts had warned Admiral Spruance that the Japanese intended at some point to unleash the *Yamato* from her berth in the Inland Sea. At 1520 on 6 April, escorted by the cruiser *Yahagi* and eight destroyers, she left Tokuyama Bay on a mission as inevitably final as that of any kamikaze.

The *Yamato* was by far the largest battleship in the world. The surviving sister ship of the *Musashi*, sunk by carrier pilots in the Sibuyan Sea during the Battle of Leyte Gulf, she dwarfed everything else afloat. Her displacement of 72,800 tons compared with our *Missouri*'s 45,000. She had 100 antiaircraft guns of various calibers, and her main battery consisted of nine 18.1 inch guns, larger than those mounted on any other ship in the world. Despite a draft of 35 feet and armor plate 16.1 inches thick on her sides and 7.87 inches thick on her deck, she could make 27 knots. Her crew numbered 3,000.

The Ten-Go Plan called for this magnificent ship to offer herself and her escorts as bait to draw Task Force 58 away from Okinawa long enough to permit the second day's strikes of Kikusui No. 1 to reach the American ships standing off the landing area. If she could fight her way past TF 58, she was to proceed to Okinawa and shoot up the newly captured airfields and as many ships of the expeditionary force as possible before she went down herself. Underscoring the hopelessness of her mission and the severity of Japan's petroleum shortage, she was fueled only for a one-way trip. No planes could be spared from the attack force to provide her with an air umbrella.

Her midshipmen had been ordered ashore that they might survive for future duty. Commanded by Vice Admiral Seiichi Ito and dubbed the First Diversionary Attack Force, this group of ten ships was the Okinawan parallel to Admiral Ozawa's four-carrier Northern Force off Cape Engaño in the Philippines.

Two American submarines, the *Threadfin* and *Hackleback*, had been stationed at the mouth of Bungo Strait to detect any traffic coming out of the Inland Sea. At 1745 on 6 April, they got off a message alerting Nimitz and Spruance that the force was transiting the strait. Almost immediately, Spruance ordered Admiral Morton L. Deyo's Okinawa Bombardment Force, composed of the older battleships and cruisers, to prepare an intercept plan; and Admiral Mitscher turned Task Force 58 to the north in the hope of finding the enemy ships and striking from the air first thing in the morning.

By dawn on the seventh, TF 58 was east of the northern tip of Okinawa and the enemy force had just rounded the southern tip of Kyushu and was headed west to draw off the carriers. Search planes fanned out from the task force and, while TG 58.4 undertook to provide the early CAPs and RAPCAPs, all the carriers made preparations to load their attack planes with torpedoes and the armor-piercing bombs.

At 0832 a search plane from the *Essex* made the first sighting. In the *Langley*'s CIC, I heard the pilot report by voice radio the latitude and longitude of his sighting, the composition of the force, which I took down as one battleship, eight destroyers, and three light carriers, and its course and speed: 300 degrees, 12 knots. Not wanting to entrust all this to our sound-powered line to the bridge or to our "squawk box," I grabbed an area chart, plotted the coordinates of the sighting, and ran up to the bridge. In every carrier, as in ours, there was intense excitement as we awaited word from the flagship as to what planes would be sent out to attack the *Yamato* and her escorts and when.

Two PBMs were sent up to shadow the group discovered by the shorter-range *Essex* search plane. These trackers did a fine job of ducking in and out of the clouds, staying just out of antiaircraft range, and reporting for hours on the movements of the enemy force. Undoubtedly, they corrected the initial report, but I did not hear their transmission. The enemy force did not include the three CVLs first reported, but did include the cruiser *Yahagi*, and it was traveling at 22 knots. It was steaming in a circular antiaircraft disposition, not unlike our own, with the *Yahagi* in the lead, the *Yamato* in the center, and the eight destroyers in a ring around her. While these ships started out with some air cover from Kyushu, the planes were withdrawn around ten o'clock and no replacements arrived.

At about the same time, the first and largest of the American strike groups took off from the eight carriers in TGs 58.1 and 58.3; fighters, dive-bombers and torpedo planes, 386 in all. A considerably smaller strike group from TG 58.4, which was heavily burdened with CAP and RAPCAP duty, followed an hour or so later with 109 planes. Commander Ed Konrad, who was Air Group Commander in the *Langley* when she was commissioned in 1943 and was now Commander Air Group 17 in the *Hornet*, served as strike commander, coordinating the attacks over the target.

The planes from TG 58.1 (the *Hornet, Bennington, Belleau Wood*, and *San Jacinto*) went in first through clouds and heavy overcast. Bomb-carrying fighters led the way, followed closely by dive-bombers and torpedo planes coming in low underneath. At 1241, two 1,000-pound bombs hit the *Yamato* and a deep-running torpedo struck her port side four minutes later. A few minutes earlier, the *Yahagi* had been hit by a torpedo and a bomb, and was dead in the water. The destroyers broke formation, and one was torpedoed and sunk in the opening minutes of the battle.

The next attack followed immediately and came from TG 58.3 (the *Essex, Bataan,* and *Bunker Hill*), whose planes had been waiting overhead for their turn. Among this group, Torpedo Squadron 84 from the *Bunker Hill*, led by Lieutenant Commander Chandler W. Swanson, was particularly effective. Determined to get the *Yamato*, and taking care that the depth settings on their torpedoes were sufficient to get in under her armor plate, these planes scored five hits on her port side, causing her to list increasingly as more and more of her compartments began to flood. A little later she was hit be three torpedoes on her port side and one to starboard.

At 1345 in a coordinated attack, torpedo planes scored seven hits and dive-bombers twelve on the *Yahagi*. Within half an hour, she went down. Two of the destroyers were severely damaged in the course of the melee.

At 1400, the strike from TG 58.4 arrived on the scene to help deliver the coup de grâce. By 1420, the *Yamato* had gone under, more than 200 miles short of the beachhead she had hoped to attack. Four of the destroyers were sunk, and two were severely damaged but able to limp off toward Kyushu. As TG 58.4's strike headed home, they saw the remaining two picking up survivors. Admiral Deyo's battleships and cruisers had no work to do, as a result of the sortie of the First Diversionary Attack Force!

The cost to TF 58 was the loss of twelve airmen and ten aircraft.

While the strike planes were carrying out their mission, the combat air patrols over the fast carriers and those over the Delegate Base pickets were busy all day long dealing with the final attacks of Kikusui No. 1. The fighters that might have been covering the Japanese ships were being used to escort 41 kamikazes on various forays against the fast carriers, while a strong force of Army aircraft attacked the ships at Okinawa. Thanks to a series of successful intercepts throughout the day, only one ship belonging to TF 58, the *Hancock*, was hit, while, closer inshore, one picket destroyer and one of Admiral Deyo's battleships took hits. The *Hancock* was forced to retire to Ulithi, compounding the trouble she suffered when her part of the first strike missed the target in the bad weather and had to return without a score.

On 8 April, Task Group 58.2 returned to the task force and we all went back to our "normal" duties off Okinawa. A typical day's operation called for all ships to be at general quarters, manning all battle stations, one hour before sunrise, and to launch the first CAP and RAPCAP forty-five minutes before sunrise. Generally, four combat air patrols were launched from a given task group during the day and each would be airborne for more than three hours. Whatever strikes or fighter sweeps were scheduled for the day would be timed for launching or landing at the same times as the CAPs were being handled on the flight decks. All ships routinely went to general quarters again and the last daylight CAP was recovered one hour before dark. In each task group one carrier equipped with night fighters would take on night duty as ordered, keeping its flight deck cleared for either night intercept or intruder missions. In CIC we kept a minimum of a Condition Three watch twenty-four hours a day, and I left standing orders that I was to be called at any time of day or night when there was a bogey on the screen. Even when the nights were quiet, we had plenty to do helping the officer of the deck keep station on the guide ship with our SG surface-search radar. No matter how dark the night, every ship in each task group zigged and zagged at intervals prescribed by a precise plan, in order to make it more difficult for enemy submarines to find us. We followed our neighboring ships very carefully on the radar to make sure everyone was turning together. At 18 knots on a dark night, one ship out of synchronization could cause a calamity.

It was at about this time in the operation that Admiral Radford, commander of TG 58.4 ordered into operation an idea that must have looked good on paper on the flag bridge, but was counterproductive in CIC. He ordered all the carriers in his group to station their combat air patrols in an orbit in a YE Sector* several miles away from the group instead of directly overhead, as had always been the practice. The purpose, of course, was to get the jump on attackers that suddenly appeared on the radar at short ranges and presumably low altitudes. In practice, however, it complicated an already too-complex radar picture. As long as the CAP was "anchored" in orbit overhead, we did not have to keep track of it until we sent it out on a vector, and it in turn

---

*YE SECTOR was one of four quadrants into which a different letter was broadcast daily in Morse Code to aid in homing lost planes.

had an easy time keeping station in any weather other than solid overcast. But when it was ordered to take station ten miles away and to stay in tight orbit, it had no reference point to guide it; and with the whole task group continually zigzagging and frequently turning into the wind to launch or land aircraft, it is no wonder that it would frequently drift or wander off its prescribed station. Under this system, the fighter directors in the carriers that had provided the CAP were responsible for policing it, and the group FDO "policed the policemen" extremely rigidly, obviously on direct orders from the admiral. The fighter director radio network crackled with transmissions all day, herding the CAP into two- and three-mile shifts of position and responding to caustic comments from the flagship to the effect that it was again off station. This was a constant irritation to the fighters as well as to the FDOs; but that was negligible compared to the problems it created with the radar. The more cluttered with targets a radar screen became, the more difficult it was to discover and assess any new target instantly. We already had to follow too many different flights of our own planes and those of adjoining task groups. Additionally, because the station-keeping of the CAP was being watched so meticulously in the flagship, we had to assign our SC radar almost full time to keeping tabs on it, whereas this radar, along with our SK, should have been searching for the enemy. There was also a big worry in my mind that a kamikaze could sneak in at low altitude on the opposite end of the group and strike before we could get the CAP back in time to intercept.

As the days wore on, many tempers grew short over this whole issue, and I sensed that Lieutenant Commander Carl Ballinger, our group FDO in Admiral Radford's flagship, understood the difficulty but could not get the order rescinded. Sure enough, a few days later he called for an unprecedented radio conference among the senior fighter director officers in all four of the carriers in the group. He alerted us during the day and set the time of the conference for that evening after all planes had landed, explaining that he wanted each of us to be prepared to give him our reaction to the new plan. I welcomed the opportunity, and prepared my objections to it carefully and as objectively as I could. While I was perhaps the most outspoken in urging a return to the overhead station, all the others were, to some degree, in favor of it, and I think we gave Ballinger good ammunition to use with the admiral's staff. The order was not rescinded, but the over-tight monitoring of it was relaxed, and we were ultimately able to put our SC radar back to work looking for the enemy.

On 11 April, Kikusui No. 2 made its appearance. Of the ten Kikusui attacks that developed before Okinawa surrendered, this one, involving 185 kamikazes escorted by 150 fighters and 45 torpedo bombers, was the second largest.

The first day's action was aimed principally at the fast carriers. Although we were able to intercept most of what came at us, we had our share of leaks; and the *Essex, Enterprise,* and *Missouri,* as well as three of our own pickets, all suffered some damage. In the course of the morning's action, one of the other carriers in our group was using a four-plane division of the CAP to intercept a twin-engined Betty, when it went into a radar fade or became lost in the melange of friendly and bogey targets around it. At any rate, the carrier reported to the group FDO that she had lost track of it. We immediately got on the air to say that we had "good dope on it," and the group FDO instantly ordered us to take over control. Every second was precious as, by then, the Betty was very close. I gave the CAP an instant vector and luck was with us: it got the Betty in sight in a minute and a half and "splashed" it in three minutes. We had been in this situation several times before, where the intercepting carrier for some reason lost a target and another, with better information, had to take over. But never before had we had such gracious recognition for doing it successfully. By blinker from Admiral Radford that afternoon came the following message, addressed to the *Langley*:

"I have just been informed that it was your FDO and CIC took over and successfully completed the interception this morning when it looked very likely the bandit would get away. A very well done to LANGLEY for this and other similar jobs. You must have a very fine team."

Needless to say, I posted my copy of this in CIC where all the radar operators and watch officers could see it. It gave us all a welcome boost, and I'm sure it did us no harm in the captain's cabin.

During the next two days, the attackers shifted to the Delegate Base pickets and other ships off the beach, including those of Admiral Deyo's bombardment group. On one of the picket stations the destroyer *Abele* was first crashed by a kamikaze and then, while dead in the water, struck by a baka. She sank immediately. In the three days that Kikusui No. 2 lasted, 14 ships were damaged and the *Abele* was lost, but enemy losses were also heavy. TF 51 claimed 147 enemy aircraft shot down; TF 58 claimed 151.

I had been up all night on the twelfth, and went below to get some sleep after the ship secured from general quarters at sunrise on the thirteenth. When I got up a couple of hours later, I went up to the flight deck and noticed that the ship's ensign was far from being two-blocked: that is to say it had not been run all the way to the top of the mast. All of us who stood officer-of-the-deck watches either underway or in port made it a habit to tip off whoever had the duty when we noticed something that was bound to bring down the wrath of the captain on the OOD. I called the bridge on the telephone from CIC and my good friend who was standing watch said: "Thanks, John, but where the hell have you been? You're the last to know that the President died, so naturally we're at half-mast!" I must have slept through the bullhorn announcement to the crew earlier in the morning.

On 14 April, the *Yorktown* was furnishing our CAP, and we directed them three times in quick succession to intercept Bettys appearing first on the *Langley*'s radar. All three intercepts were successful. One of the Bettys was carrying a baka. The two following days, we "went both ways at once," supplying strikes and CAP to Okinawa and sending strong fighter sweeps against airfields in southern Kyushu. The latter had good hunting and reported destroying 80 planes in the air and on the ground, preventing their being added to the 165 kamikazes of Kikusui No. 3, which attacked on the sixteenth.

This one divided its attention between the Delegate Base pickets, where they scored heavily, and TF 58 where a kamikaze crashed into the flight deck of the *Intrepid*, not only putting her out of action, but sending her all the way to the United States for major repairs. Ironically, we successfully directed the *Intrepid*'s fighters in shooting down a Kate 20 miles away, and a Zero 30 miles away. The destroyer *Pringle* was sunk by a kamikaze on Picket Station 14, and nine ships were damaged.

After this action, TG 58.2 was dissolved and its remaining carriers, the *Randolph* and *Independence* were assigned to other groups.

In addition to the baka bomb, kamikaze, two-man submarine, and suicide boat, the Japanese had in their arsenal of desperation the kaiten. This was a long-range torpedo carrying 3,200 pounds of explosive, capable of running up to 83 miles and, on shorter runs, achieving a speed of 40 knots. It was launched from a mother submarine, and directed into its target by a human pilot.

On 18 April, five destroyers from our screen teamed up to sink the submarine I-56, which was carrying a litter of kaiten. Shortly after midnight, the destroyer *Uhlmann* made the first attack, dropping two full patterns of depth charges over a strong sonar contact. The *Herman, McCord, Collette,* and *Mertz* came up to help finish the job.

While most of the work in CIC was of necessity grim, we did have one subspecialty that we all enjoyed: finding and homing lost pilots. I remember with pleasure two examples from the *Langley*'s experiences off Okinawa.

Early in the operation, the airfields on the island were secured for US use and Marine fighter squadrons flew their F4Us ashore and based them there. When they were assigned to CAP duty over a picket destroyer at one of the more distant stations, such as No. 8 or No. 14, and when they found themselves in hot pursuit of enemy planes, it was very easy for them to lose track of where they were. Late one afternoon, while we were operating well to the east of the island, I noticed a group of aircraft circling around, some 30 miles to the northeast. They were clearly showing IFF, so I knew they were friendly, and I assumed they were looking for something on the water. But when time went by and they were still in the same general area, I sent a division from the CAP out to look them over. Finding them to be F4Us, our fighters got through to them on whatever radio channel they were using and then called us to say that they were bringing us four Marines who were lost and very low on gas. On the way back, it developed that the Marines had never landed on a carrier. Our deck was smaller and more difficult than the decks of the *Essex* class, but, since only ours was spotted for landing, it was ours or the drink.

We got hold of the air officer immediately and he came to CIC to see what we could do. Hastily we decided to plug in a radio circuit way aft, at the landing signal officer's platform, so that he and the air officer could coach them in. These arrangements were then relayed to the incoming group which soon arrived. We had meanwhile reported the circumstances to the flagship, and when the planes came in sight the whole task group turned into the wind. In the calmest possible tones our LSO explained the procedures to the Marines, reassured them without overdoing it, and directed our own planes to lead them down into the landing circle. Two of the F4Us made it safely into the arresting gear without a wave-off; one needed two passes, and the last needed three. They were down to their very last few gallons of gas, and I think all of us involved breathed almost as big a sigh of relief as they did!

Another time, we heard a plane belonging to another task group calling his carrier in some distress. After calling repeatedly without being able to raise his "base," he asked any ship in his task group to respond. None did. He finally asked *any* ship that heard him to reply. We heard him clearly and I called him immediately. He had become separated from his strike group over Okinawa, was unsure of either his own or his carrier's position, and was trying to get back. There must have been something wrong with his radio crystal for his own group's frequency, so he had been trying to reach them on ours. First, we asked him to turn on his "emergency lights," fighter director's code for the emergency switch on his IFF, which gave off a distinctive vertical flash allowing us to spot him from among all the other planes on the radar screen. We thought we saw him right away, and plotted his course and speed from two other plots one minute apart; we then asked him if he was on that course and at that speed. When he gave us a relieved affirmative, we were sure we had him and told him to turn off his "emergency lights." When they promptly went off on the PPI scope in front of me, I told him we had him for sure, asked him how much gas he had left, and said we would give him directions in a moment. Meanwhile, we had tried to get his task group on our SG, but they were beyond our range over the horizon. However, we had a plot on that group's CAP orbiting over it. He only had about 25 miles to go and reported enough gas to get there, with some to spare. We then gave him a slight alteration to the course he had been flying, told him we thought he would see his base in five minutes, and asked him to call us when he did. Sure enough, he saw it right on schedule, and called us with such relief in his voice that we all felt warmed by it. Scary, indeed, to be lost over the ocean, low on gas, lonely and unable to reach anyone "at home."

While the land battle on Okinawa became more and more bitter, the attacks on the fast carriers diminished perceptibly, as the enemy rounded up enough kamikazes for Kikusui No. 4. They unleashed this one, which had 115 kamikazes committed, on 27 and 28 April. They sank an ammunition ship and damaged nine other ships off the beach and on the picket stations. The hospital ship *Comfort* was crashed at night while proceeding, fully lighted, to Saipan. Her entire medical corps was wiped out and many nurses and patients were killed or wounded. On the twenty-ninth, the remnants of the attack force struck at the fast carriers and two destroyers, the *Haggard* and *Hazelwood*; both from the screen in our task group, were severely damaged.

By the end of April, there was no question that both sides were hurting badly. In the Kikusui attacks alone, 7 US ships had been sunk and 56 damaged, and many more had been put out of action for various periods by the smaller, but no less vicious attacks by kamikazes in the intervals between Kikusui. It was obvious that, for the troops on the island, the struggle would continue to the last Japanese, and the end was still some way off. On the other hand, Japanese aircraft and pilot losses in the Kikusui attacks added up to 820 during the month of April, and that did not count heavy losses among their escorts, planes and pilots lost in the intervals between the Kikusui attacks, and the many aircraft destroyed in the air and on the ground by TF 58's raids on Kyushu, Shikoku, Honshu, Minami, and Amani. The British carriers' constant pressure on the Sakishima Gunto accounted for many more. Taken together with the loss of the *Yamato* force, it was clear that something in Japan would soon give way. It seemed to us that there just couldn't be many trained pilots and serviceable aircraft left in the Empire.

Having been at sea continuously for a month and a half, TF 58 was in need of maintenance work that couldn't be done underway. TG 58.1 was therefore sent back to Ulithi for ten days, while TG 58.3 and TG 58.4 remained on station.

Fortunately for all of us, there were bits of humor here and there to relieve the stress.

On a quiet refueling day, one of our pilots on combat air patrol got bored with orbiting indefinitely over the ship with no bogeys on the screen, and used the button of his microphone to transmit the following in Morse code:

```
 ..-      ...      ...
 ..-   -..-.   ....    .    ...
```

In CIC, I turned to Tom Sorber and said, "At flight quarters tomorrow, let's issue him a pillow instead of a parachute!"

Early in April, I myself violated radio procedure in an attempt to be funny. We had just completed a successful intercept on a twin-engined plane, a Betty, that was flying all by itself at 19,000 feet on a beeline from Formosa toward Japan. When I reported the "splash" to the flagship, the watch officer in the *Yorktown*'s CIC acknowledged and then asked me what I thought the plane was doing. Obviously I had no way of knowing, so I answered, "Dunno, maybe just a Vulture from Culture." Since "Culture" was the code name for Formosa, I assumed everybody would realize this was just a play on the old expression "Vulture for Culture," but a long time later an FDO from another ship said to me, "I know where Culture is, but what in hell does Vulture mean?"

In the *Langley*, we had a twice daily diversion during general quarters. We had on board an extremely nervous assistant air officer, new to the ship. This individual's battle station was in air plot, adjacent to us and connected by an open doorway, but he had no duty to perform there. In his anxiety and nervousness, he would wander back and forth between air plot and radar plot for the whole hour, morning and evening. At first this was just annoying, but soon we got up a pool of how many times he would come in during a single hour's GQ. One day, when I put my quarter down on 31, I hit it on the nose and won a considerable sum.

Every once in a while, some ship would make a pass at "The Galloping Ghost of Nansei Shoto." It showed itself as a radar blip, not quite as sharply defined as an airplane's, sometimes moving slowly as if in orbit. It appeared only once in about every couple of weeks, and then someone was bound to report it as a bogey. CAPs "intercepted" it, finding nothing but thin air; destroyers sometimes fired at it; and as far as I know, no one ever figured out what caused it. All I know is that it never showed IFF.

On 3 May the enemy unleashed Kikusui No. 5, an especially effective attack on the Delegate Base pickets, sinking two destroyers and two LSMs, and damaging fourteen other vessels. Not since the first attack in April had they scored that high, nor were they to do so again.

During a day of hectic action, we took part in a curious radar anomoly. While steaming well to the southeast of Okinawa, we detected a huge bogey 120 miles to the north. Only once before, off Saipan, had we detected a target that far away; the normal limit of our radar, no matter how high a target was flying, was approximately 80 miles. When we reported this contact to the group FDO, he asked for its course and speed, which we computed after two more plots. It was a very large raid, on a course that would take it not toward us but straight to Okinawa. The group FDO passed on this information to Delegate Base, and apparently neither they nor the pickets to the northeast of the island yet had the target on their screens: nor did the *Yorktown* or even the neighboring task group, which was 16 miles closer to the target. In some excitement, the group FDO asked us to give him a plot every minute, which he in turn relayed to Delegate, who started fighters out to intercept. It was several minutes before this raid appeared on Delegate's own screens, even though he was many miles closer to it than we. Eventually, everyone had it, but we could never explain to ourselves why we saw it when it was so far away. We just hoped that the extra few minutes' warning we provided saved some lives on the picket line.

On 4 May we had a second very pleasant surprise from the flagship. Commander J.J. Southerland, commanding our Air Group 23 as replacement for the grievous loss of two former commanders of the group, returned from a visit to the *Yorktown* with the following handwritten note, to which I have added only the material in brackets:

"LANGLEY CIC:

"I'm somewhat ashamed that I have not sent a note over before to tell you what we think of you although you may know.

"In short we have enjoyed working with you as much as this type of life can be enjoyed and we think you're the top team. I, myself, have needed an understanding answer from the group more than once and you usually furnish it. Not to forget the chestnuts I have to ask you to pull out of the fire now and then . . . a hell of a reward for being reliable.

"The YORKTOWN CIC has a bellyful of being a flagship and my ulcer is in foul shape from arguments with guess who [obviously Admiral Radford] about stationing the CAP and air operation. Fortunately none of this has affected the group in the ultimate result which is a good job all around. Sorry the Oporder fell apart . . . writing a new one . . . no wonder the S-L [presumably, the *St. Louis*, a cruiser in TG 58.4] is having a hard time with nothing in writing to go on.

"We're looking forward to that forthcoming rest [rotation of groups to Ulithi] and perhaps a chance to meet you at Mog-Mog with a beer in each hand. I hope we can get some more planes. Looks like we may tomorrow. The beauty of knocking them off here is that it means the end of a pilot too.

"Joe Eggert [FDO on Admiral Mitscher's staff] sent a note in a happy mood . . . I can remember our betting CTF 58 [Admiral Mitscher] that the FD teams could stop more than 90% of the Japs with the CAP. Evidently CTF thinks the job is OK and told Joe that. Thank God for the Pick line.

"Looks like a clear screen tomorrow so if they come we can show a clean performance.

"Cordially, Carl Ballinger"

This frank and appreciative note did a lot to cheer things up, and I circulated it to each watch officer in CIC.

The news that Germany had surrendered was flashed throughout the force on 7 May. We were happy for everyone concerned, and hoped that the news would tip the scale and make the Japanese realize that the time had come to quit.

After refueling, we were back on the line on 11 May when Kikusui No. 6 struck with another 150 kamikazes, some going for the pickets and some for us. In the course of that day's action, at 1015, the *Langley*'s fighters intercepted a Myrt at 21,000 feet. An hour later, the *Shangri La*'s fighters under our direction splashed a Jill 40 miles to the east, and the *Langley*'s fighters shot down three Zekes and a Jill 20 miles to the northeast. Admiral Mitscher's flagship, the *Bunker Hill*, took two kamikazes in her flight deck and was so badly hurt that she had to return to the West Coast. He shifted his flag to the *Enterprise* which got it herself a few days hence, requiring the admiral to shift once again, to the *Randolph*.

In the *Langley*, when our last strike of the day came back aboard, we received Admiral Mitscher's order to our task group to head for Ulithi for maintenance and repairs. En route, the *Langley* received a dispatch from CinCPac ordering that she be detached from the task force on arrival at Ulithi, and proceed, via Pearl Harbor, to the West Coast for the major overhaul that had been long, long delayed.

To summarize very briefly the subsequent work of the fast carrier task force against the kamikazes in support of Okinawa:

...Groups 58.1 and 58.3 hit hard at the home islands immediately after our departure.

...The *Enterprise* was hit in a 26-plane raid on 14 May and forced to return to the states.

...Kikusui No. 7 and No. 8 struck against the pickets and transports during the last week in May, sinking one ship between them. No. 9 and No. 10 which struck in June, consisted of only 50 and 45 kamikazes, respectively, and did not sink anything.

...Admiral Halsey relieved Admiral Spruance on 27 May and TF 58 became TF 38.

...On 4 June the task force was battered in a typhoon and, after recovering sufficiently to get off a few strikes and combat air patrols, retired for repairs to Leyte, arriving there on 13 June. Two weeks later, it was ready to sail again and, instead of going back to Okinawa, was at last unleashed to spend the rest of the war striking at the Japanese home islands.

In his final report on the war, Admiral King tells us that the fast carriers destroyed 2,336 enemy planes in support of Okinawa, at a cost of 557 of their own.* Admiral Morison reports that the Navy lost 30 ships sunk, mostly by kamikazes, and 368 damaged, both including some craft smaller than destroyers.** Commander Nakajima lists a total of 1,465 kamikazes expended in the ten Kikusui attacks and a grand total of "almost 1,700" kamikazes during the Okinawa operation.*** Additionally, TF 58 sank the largest battleship left in the world, one of the last of Japan's cruisers, and four destroyers.

Nearly forty years later, when I think of Okinawa, I think of the incredible ordeal of our sailors manning the ships on the radar picket line; and wonder why in the world the Japanese expended so many hundreds of their kamikazes on those lonely destroyers and LSMs instead of sticking to their original Ten-Go plan for concentrating on the carriers and transports.

And I marvel still that so many of us in the *Langley* were able to survive. For to us, Operation Iceberg was not something that could be neatly compartmentalized, as is an operation plan or a history book. We fought it as the climax of a continuum that began with our departure from San Diego in December 1943 and included, without any real respite, the capture of Kwajalein and Eniwetok in the Marshalls; the landings at Hollandia on New Guinea; the strikes on Truk; the taking of Saipan, Tinian, and Guam in the Marianas and the Battle of the Philippine Sea; the capture of Peleliu in the Palaus; the entire Philippine campaign from Mindinao to Luzon, including the massive Battle for Leyte Gulf; the strikes on Formosa and the foray into the South China Sea to the ports of Indochina and China; the incredible typhoon in December 1944; the hit from a kamikaze when we were close to the shore of Formosa; the assault on Iwo Jima and the raids on Tokyo; and, finally, all that came our way off Okinawa.

Although there is no way to measure it, there is no question that fatigue played an important role in the performance of our forces in the Navy. The Navy was not insensible to the problem and set up rotational schemes for ships, squadrons, and personnel. Once the Navy pilot-training program got into high gear, replacement pilots were available at every refueling rendezvous; and carrier air groups could be rotated at reasonable intervals, according to the severity of their operations.

---

*Ernest J. King, *U.S. Navy at War 1941-1945* (Washington, DC: US Navy Department, 1946), p. 14.
**Morison, p. 282.
***Inoguchi et al., pp. 151 and 160.

The *Langley* had three separate air groups during her battle cruise. Carrier rotation was something else. Whatever the Navy intended as the rotation interval for such ships (some said six months), the actual interval was wholly unpredictable when the carriers became the prime targets of the kamikazes. So many carriers were hit in the course of operations that those which were not had to stay on, sometimes long past reasonable time limits. JOMO (Just One More Operation) became a common and cynical acronym on board. With the added stresses of tight watch schedules, countless numbers of raids, day and night, and long hours at general quarters, serving in the carriers became a severe test of endurance that not everyone could handle.

If one adds to all the above heavy stresses that came with the extra burden of command, it would be both unrealistic and ungenerous to deny a carrier captain the occasional luxury of some pretty odd behavior.

XXII
Homeward Bound

The *Langley* was well on her way to Ulithi when her captain received orders to return to Pearl Harbor and thence to San Francisco for major overhaul. Under other circumstances, the ship would have been sent back at least six months sooner, but because so many carriers were put out of action by the kamikazes, the *Langley* had had to continue for many additional operations. There had been little rotation among the ship's crew and the men were exhausted. We had sailed her 160,000 miles, most of which were "in harm's way." Given these circumstances, I thought the captain could have chosen a better way to announce to the crew our long-awaited orders. The way he did it was to have a boatswain's mate call over the loudspeaker system: "Now hear this.... Any man without a haircut will not be allowed to go ashore at Pearl Harbor!"

At Ulithi, we received a dispatch detaching me from the ship. Translated from the Navy cablese, it read: "From: ComAirPac to: USS *Langley*. Lieutenant John None Monsarrat hereby detached. Proceed and report to the Pacific Fleet Radar Center by first available air transportation for further assignment by Commander, Service Force, Pacific...."

Excited as I was at the prospect of going home, I did not want to be stuck in Ulithi waiting for questionable air transportation while the ship sailed off without me. So Tom Smith kindly sent a message to the Marine Transport Air Group (TAG) on the beach: he referred to my orders and asked how long it might be before they could find a place for me on a plane. In short order, back came the following reply, "Air transportation very indefinite at present. Suggest taking surface if available. TAG."

Here, I thought were all the elements of a classic Navy mixup, such as the one that found Tom Smith "lost" in Espiritu Santo for six months after his ship was sunk. So, armed with the TAG dispatch, I went to the captain's cabin and asked his orderly if I could see him. When I was admitted, I said to the captain: "Sir, I have orders to return to Pearl by first available air, but the Transport Air Group on the beach advises me to go by surface if at all possible. Since the ship is now going back, may I have permission to go back in her, rather than sit in Ulithi waiting for air transportation to open up?"

After looking at my dispatches, the captain replied, "Monsarrat, your orders read 'first available *air* transportation,' and that's the way you're going to go," his tone seeming to imply "even if it takes all summer!"

After packing up my gear and saying goodbye to as many shipmates as I could, I took a whaleboat into the beach and found the duty officer's office at the Marine Air Base. I explained that although we had received TAG's suggestion that I go by surface, our captain wouldn't buy it, and I would appreciate whatever TAG could work out for me. A sympathetic Marine captain consulted an operations clipboard for a few moments, then said: "We have a C-46 going to Guam this afternoon. I can put you on it if you want me to, but you may have a long wait in Guam before you can get a flight to Pearl. They're loaded with Okinawan casualties there, and naturally they get the top priorities for eastbound flights."

I didn't hesitate; Guam was in the right direction. "Just tell me where and when to report aboard," I replied.

So it was that as the *Langley* pulled out of the lagoon, I took off for Guam, some 300 miles closer to home. I shared an Army tent with other transients until, a week later, a seat opened up on a flight for Pearl. By an ironic coincidence the day I landed at Hickam Field, Pearl Harbor, the *Langley* steamed in to tie up at Ford Island just across the harbor!

When I reported to the Fleet Radar Center at Camp Catlin, I simply didn't recognize the place. Gone were all the temporary shacks we used when I was in Fighter Director School there, and I needed a guide to show me where to go. When finally I found the right office, I produced my orders and was given a new set to proceed by air to San Francisco. After completing the formalities, I hunted up the personnel officer and we went over a list of the radarmen who had been in the *Langley* as long as I had. He was amazed that we had been out in "Indian country" so long without any real rotation, and promised to see that as many as possible of our radarmen got relieved when the ship reached San Francisco.

Bearing in mind that we had trained all our radarmen on board, I had quite a list to hand him. One of our men, Mullins, had made chief radarman and we had transferred him along with another who had made first class. But still on board were:

RdM1C	RdM2C	RdM3C	S1C	S2C
Polinsky	Loughran	Katra	Hickam	Bonner
	Roser	Kelley, J. L.		Fernandez
	Kassay	Laughlin		Eierdam
	Duffy	Hanks		Everett
	Coder	Marshall		
	Kelly, R. J.	Cook		
	Freeda	Harrison		
	Cheda	Gerhart		
	Hoover	Barnes		
	Finlay	Taylor		
	Capuro	Berry		
	North			
	Gutosky			
	Pederson			
	Clearwater			
	Hotchkiss			
	Sharwarko			

I flew to San Francisco on a four-engined flying boat, a PB2Y. We took off in the late afternoon and, as darkness settled down, a huge golden moon rose to slowly turn silver and light up the waves beneath us. It was one of the most beautiful nights I can remember, and between its beauty and my excitement, I didn't sleep a wink, and didn't even want to. I only wanted to think of home and savor the experience of heading there.

Once landed at Treasure Island, I made a beeline to the Naval District Headquarters, where I received orders for thirty days' leave and a priority for airline transportation to New York. There, I learned that the *Langley* had docked that very morning. Although I was frantic to call home, I thought I had better find out when I could get a flight so that I would be able to tell Peggy when to expect me. So I went first to the United Airlines' office near the St. Francis Hotel and waited my turn in a long line at the ticket counter. At last, I got a ticket for the next day, and found a hotel room for the night.

Peggy had left Columbus after the death of her father, and returned with Nicholas to our house on Crawford Road in Westport. I'm sure my voice was shaking when I put in the first telephone call to her in more than a year and a half. I had not even been allowed to write her that I was en route home. When my call went through from my hotel room, I can't begin to describe my emotion. Keyed up as I was, I was absolutely flabbergasted to realize that it was *Nick* who answered my call! He was then three and a half years old, but I had never heard him speak anything but baby talk, and at first I could not believe my ears. In all my life I have never felt such a wave of emotion and joy and relief at the sound of his and then his mother's voice during the next few minutes. It was as if the whole world had begun again after an interminable pause in space!

Once at home, time flew. During my leave Bill Forbes gave a dinner for me in Hartford, at which I told Bill Gwinn and others from United Aircraft as much as I could about events in the Pacific, although we were still forbidden to talk about the kamikazes. When my leave ended, I reported in New York, and was sent to St. Albans Naval Hospital for a thorough checkup before being reassigned to duty. The bomb was dropped on Hiroshima and the Japanese surrendered while I was at St. Albans. For some reason, the idea of joining the frenzied crowds in Times Square on the night of V-J Day repelled me. I just wanted to go to church, which I did as soon as I could.

Inevitably, there was delay and much red tape before I was released from active duty, but eventually the great day came. On my last day of active duty, I became eligible for promotion to lieutenant commander; and after my final papers had been signed and stamped in downtown Manhattan, I called Bill Forbes to invite him to join me in a celebration luncheon.

"Great," said Bill, "But you come and join us. We're meeting Harold Mansfield from Boeing in Seattle and Deac Lyman to talk about postwar problems in the airline industry, and it'll be a great way for you to get back to business again!"

And so it was that, still in uniform, I went back to work, putting aside chasing bogeys, the business of fighter direction, and returning to building brands, the business of advertising.

A few weeks later, in a kind of curtain call to my service in the Navy, I was summoned to the headquarters of the Naval District in New York and presented with the Navy Bronze Star Medal and the following citation:

"For meritorious achievement as Fighter Director Officer of the U.S.S. LANGLEY during operations against enemy Japanese forces in the vicinity of Kyushu and Okinawa from March 18 to May 11, 1945. Exercising skill in the performance of his duties, Lieutenant Monsarrat directed fighter planes in the successful interception of fourteen separate enemy air attacks against our forces, thereby contributing materially to the destruction of a large number of hostile aircraft at a considerable distance from the Task Group to which his ship was assigned. His tireless devotion to duty was in keeping with the highest traditions of the United States Naval Service.

For the President,
James Forrestal
Secretary of the Navy"

Chronology of John Monsarrat's Service in the US Naval Reserve

29 November 1912: Born in Columbus, Ohio.

24 June 1942: Commissioned lieutenant (j.g.), USNR.

1 July 1942: Ordered to active duty and reported to Naval Training School, Harvard University.

28 August 1942: Completed Indoctrination School at Harvard.

29 August 1942: Reported for duty in Bureau of Aeronautics, Washington, DC, as Officer-In-Charge, Editorial Research.

20 January 1943: Detached from Bureau of Aeronautics and ordered to Staff, Commander in Chief, Pacific Fleet.

16 February 1943: Reported to Commander in Chief, Pacific Fleet, at Makalapa, Oahu, and assigned to temporary duty on Staff, Commander, Air Force, Pacific Fleet, Ford Island, Pearl Harbor.

3 April 1943: Detached from Staff, Commander, Air Force, Pacific, and returned to Staff, Commander in Chief, Pacific Fleet.

28 April 1943: Ordered to temporary duty in Fleet Radar Center, Camp Catlin, Oahu, and reported to Fighter Director and Combat Information Center school.

6 June 1943: Ordered and reported to the USS *Bennett* (DD 473) for temporary duty in connection with fighter direction.

12 June 1943: Detached from the *Bennett* and returned to Fighter Director school.

20 June 1943: Promoted to lieutenant.

24 June 1943: Ordered and reported to the USS *Essex* (CV 9) for temporary duty in connection with fighter direction.

27 June 1943: Detached from the *Essex* and returned to Fighter Director school, and carried out verbal orders to the HMS *Victorious* to observe Royal Navy fighter direction exercise.

3 July 1943: Detached from Pacific Fleet Radar Center and CinCPac Staff, and ordered to Precommissioning Detail, USS *Langley* (CVL 27) in Camden, NJ.
27 July 1943: Reported to precommissioning detail.
9 August 1943: Reported to VF 32 in Atlantic City for duty in fighter direction with squadron due to serve aboard the *Langley*.
17 August 1943: Reported to Beavertail Point, Jamestown, RI, to continue exercises with VF 32, then at Quonset Point.
24 August 1943: Returned to the *Langley* in Philadelphia Navy Yard.
31 August 1943: USS *Langley* commissioned, Philadelphia Navy Yard.
1 October–22 November 1943: Shakedown cruise to Gulf of Paria, Venezuela.
6 December 1943: Departed Philadelphia for San Diego and Pearl Harbor.
24 December 1943: Arrived Pearl Harbor.
25 December 1943: Ordered and reported to the USS *Yorktown* (CV 10) for temporary duty in fighter direction.
1 January 1944: Returned to the *Langley* at Pearl Harbor.
January 1944–May 1945: 24 successive combat operations aboard the *Langley*.
16 May 1945: Detached from the *Langley* in Ulithi Lagoon and proceeded to Third Naval District Headquarters, New York City.
13 July 1945: Reported to St. Albans Naval Hospital for examination.
21 September 1945: Discharged from St. Albans and reported for temporary duty to USN Separation Center, Lido Beach, NY.
12 October 1945: Released from active duty.
5 October 1948: Promoted to lieutenant commander to rank from 3 October 1945.
1 January 1954: Retired from Naval Reserve and promoted to commander.

Decorations:
Bronze Star Medal with Combat V.
Navy Commendation Medal with Combat V.
Navy Unit Commendation.
American Campaign Medal.
Asiatic-Pacific Campaign Medal with nine battle stars.
World War II Victory Medal.
Philippine Liberation Medal with two bronze stars.
Philippine Presidential Unit Citation.

Appendix
Log of the USS *Langley* (CVL-27)
November 1943 to July 1945

November 1943
- Departed Philadelphia, called at Norfolk, proceeded to Trinidad and returned to Philadelphia (shakedown cruise).

December 1943
- 6 Departed Philadelphia.
- 11 Through Panama Canal.
- 20 Arrive San Diego.
- 21 Departed San Diego. Marines and aircraft aboard.
- 25 Arrive Pearl Harbor.

January 1944
- 19 Departed Pearl Harbor—Task Force 58.
- 28 East of Marshall Islands.
- 29-31 Air attacks on Wotje, Taroa.

February 1944
- 1-29 Air attacks on Engebi, Eniwetok, Kwajalein, Parry atolls. Month of February continued bombing of atolls. Kwajalein and Eniwetok used as anchorages.

March 1944
- 1-7 Anchored at Majuro—Task Force 58.
- 9 Departed Majuro atoll. Crossed equator 1900 hours 172 degrees east longitude.
- 13 Arrived and anchored Espirito Santo, New Hebrides.
- 23 Departed New Hebrides, course northwest.
- 27 Rendezvous with two other task groups north of Solomon Islands. Task Force 58 now 3 task groups. Quite a sight.
- 28-30 South of Palau. Jap air attacks both days.
- 31 Air attacks on Palau.

April 1944
- 1 Air attacks on Woleai Island.
- 2 Course east.
- 6 Arrive and anchor Majuro atoll, Marshall Islands.
- 13 Departed Majuro, course southwest.
- 21-23 Air attacks on Hollandia, New Guinea and supported troop landings.
- 24 Course northeast.
- 29-30 Air attack on Truk atoll. Pilots down a record of 24 aircraft. *Lexington* near bomb hit.

182   Angel on the Yardarm

May 1944
- 1 Course northeast.
- 4-31 Arrived and anchored Majuro atoll, Marshall Islands. (Sea and gunnery practice.)

June 1944
- 1-5 Anchored Majuro.
- 6 Departed Majuro, course northwest.
- 10 West of Saipan.
- 11-13 Air attacks on Saipan, Marianas Islands.
- 15-17 Air attacks Iwo Jima and Pagan, Bonin Islands. Searches for Jap Fleet west of Marianas.
- 19 West of Marianas. Jap carrier planes attack task force all day. Estimate 300 Jap planes destroyed.
- 20-21 Remain in Saipan area.
- 22 Jap planes in area of task force.
- 23-30 Air strikes on Pagan and Rota Islands.

July 1944
- 1-2 Air strikes on Rota Island.
- 3 Departed Marianas Islands area.
- 6 Anchored Eniwetok.
- 14 Depart Eniwetok, course west.
- 18-31 Air strikes on Guam.

August 1944
- 1-9 Air strikes on Guam.
- 10 Depart Guam area, course southeast.
- 13-28 Anchored Eniwetok.
- 29 Depart Eniwetok, course southwest.

September 1944
- 1-5 Operating south of Truk, north of the Solomons.
- 6-8 Air strikes on Palau.
- 9-12 Air strikes on Mindanao and Cebu Islands in the Philippines.
- 13 Air strikes on Cebu, close bomb hit off port bow.
- 14 Air strikes on Panay Island.
- 15-20 Operating south of Palau.
- 21-24 Air strikes in the Manila area, Philippine Islands.
- 25 Depart Philippine Island area.
- 27-30 Operate from Palau anchorage.

October 1944
- 1 Depart Palau.
- 2-5 Anchored Ulithi atoll. Air Group 44 relieves Air Group 32.
- 6 Depart Ulithi atoll, course northwest.
- 10-11 Air strikes on Nansai Shoto Islands southwest of Japan.
- 12-14 Air strikes on Formosa. Jap air attacks. *Langley* downs one aircraft.
- 15-23 East of Philippines. Search for Jap Fleet. Air attacks on Manila.
- 24 Northeast of Luzon. Jap air attacks. *Princeton* hit badly by dive bomber. *Langley* gunners down three enemy aircraft.

Appendix 183

    25   Air strikes on Japanese Fleet. No attacks our task force. Damage to enemy fleet.
    26   Depart area, course east.
30-31   Anchored Ulithi atoll.

November 1944
    1   Depart Ulithi, course northwest.
    5   Air strikes on Luzon. Under suicide attack in a.m. *Lexington* hit off our starboard quarter.
  6-10   East of Luzon. Bad storms. Speed run west to Philippines.
11-14   East of Luzon. Attack enemy force west of Philippines. Air strikes on Manila.
    15   Depart area, course east.
17-19   Anchor Ulithi atoll.
    20   Jap midget subs in atoll torpedo one tanker.
    22   Depart Ulithi.
    25   Air strikes on Luzon. Jap air attacks. *Essex* hit.
26-30   Off Philippines.

December 1944
  2-10   Anchor Ulithi atoll.
    11   Depart Ulithi, course west.
    14   Air strikes on Luzon.
    15   D-day, cover landings on Mindoro Island.
    16   Depart area.
17-23   Severe typhoons, destroyer lost.
24-29   Anchored Ulithi atoll.
    30   Depart Ulithi.

January 1945
  3-4   Air strikes on Formosa.
  5-9   Southeast of Formosa. Air strikes on Formosa and Luzon.
10-11   Northwest of Luzon into South China Sea.
12-16   Air strikes on French Indochina, Hong Kong, harbors, convoys.
17-20   Course west to operating area north of Luzon.
    21   Air strikes on Formosa. Under air attack. *Langley* hit forward by 100# bomb. *Ticonderoga* hit by kamikaze.
    22   Air strikes on Nansai Shoto.
23-24   Underway southeast.
25-28   Anchored Ulithi atoll.

February 1945
  1-9   Repairs to ship, recreation mog-mog. Air Group 23 relieves Air Group 44.
    10   Depart Ulithi, course northeast.
16-17   Air strikes on Tokyo.
    18   Air strikes on Bonin Islands.
    21   West of Bonin, under Jap air attack. Intercept Jap planes heading for Iwo Jima.
23-24   Refuel at sea.
    25   Air raids on Tokyo.

Angel on the Yardarm

26-28	Course south, very bad seas, typhoon.

March 1945

1-10	Anchored Ulithi atoll.
11	At anchor, *Randolph* hit by kamikaze.
12-13	At anchor.
14	Depart Ulithi, course northwest.
18-19	Air strikes on Kyushu, Japan. Under air attack all day. *Enterprise* hit by bomb, *Intrepid* hit. Jap plane crashes off port quarter.
19	*Franklin* hit by kamikaze. Large fires.
20	South of Kyushu, air cover for *Franklin*.
21	Under Jap air attack.
22-29	Course southwest. Air strikes on Okinawa, search Jap Fleet south of Japan.
30-31	Air strikes on Okinawa, enemy air attacks, two *Langley* fighters shot down by task force.

April 1945

1	"Love" Day on Okinawa.
2-6	Cover landings, air strikes.
7	Off Okinawa. Our aircraft locate Jap surface force 200 miles north. Air attacks on Jap surface force, sunk one battleship, two cruisers, three destroyers.
8-10	Air cover Okinawa, refuel.
11-14	Support troops on Okinawa. Under Jap air attack, many enemy planes shot down.
15	Refuel at sea.
16-17	Intercept large Jap air raid on Okinawa, under kamikaze attack, *Intrepid* hit off starboard bow.
18-30	Continue support troops on Okinawa. Frequent day and night enemy air attacks.

May 1945

1-10	East of Okinawa supporting troops.
11	Under Jap air attack, *Bunker Hill* hit by kamikaze. News that *Langley* going home.
12	Depart operating area.
14	Anchor Ulithi atoll.
17	Depart Ulithi, course east.
25	Arrive Pearl Harbor (departed Pearl 19 January 1944).
28	Depart Pearl Harbor, course east.

June 1945

3	Arrive Alameda, California.
4-28	Leave and liberty.

Index

A

Abele destroyer 164
Air Combat Intelligence 25, 29, 30, 56
Air Group 43, 45, 139, 140
air-search radar, SC 33, 39, 40, 54, 164
Albacore submarine 90
altitude of aircraft estimates 61-62, 64
Angaur 93
Aobe 76
August, Lt. Charles 126

B

baka bomb see kamikazes
Baldwin, Hanson W., Leyte Gulf, Battle of 100; the Typhoon 121-22
Balintang Channel 127, 128
Ballinger, Cdr. Carl 163, 170
Barber, Lt. 84
Bashi Channel 127
Bates, Capt. Richard W. (Rafe) 106
Belleau Wood carrier 19, 111, 160
Belmont, Lt. August 30
Bennett destroyer 39, 112
Bilava, Lt. 36
Birmingham cruiser 104, 105, 107
Bogan, R.Adm. Gerald F. 91, 106-07, 110, 130
Bohol 94
British Naval Aviation 21, 22, 23, 37, 38
British Radar School 45
Brunmeyer, Carl 103-04
Brush and *Maddox* destroyers 130, 137, 138
Bungo Strait 159
Bunker Hill carrier 45
Buracker, Capt. 104
Bureau of Aeronautics 7, 8, 9-14, 22, 55
Butler, Lt. Cdr. Glen 113

C

Cabot carrier 47, 48, 98, 114, 119, 144
Camp Catlin (Radar School) 33-36, 38, 175
Camranh Bay 126, 127
Carr, Lt. Stuart 131, 133, 134
Cavalla submarine 90
Cebu 94
Chitose carrier 110
Chiyoda cruiser 110
Churchill, Hon. Winston S. 22, 24
Clark, R.Adm. Joseph J. (Jocko) 69, 80, 87
Clark Field airport 112
Clifton, Jumping Joe 71
Coast watchers 38-39

Colahan and *Cassin Young* destroyers 130, 133
Combat Air Patrol (CAP) 39, 40
Combat Information Center (CIC) 34
Comfort hospital ship 167
Commander Air Pacific 19, 29, 30, 36
Commander Destroyer Pacific 39
Commander in Chief, Pacific 25, 31, 42, 171
Cowpens carrier 48, 98, 119

D

Davison, Adm. Ralph 19, 106, 107-10, 152
de Gaulle, Gen. Charles 23, 24
de Weldon, Felix 147
Deyo, Adm. Morton L. 159, 164
Dillon, Capt. Wallace M. (Gotch) 55, 68, 78-9, 95, 113
Doughty, Lt. Cdr. Morris R. (Dutch) 45, 47-8, 59, 60, 68, 69, 96, 104, 107, 113, 124, 131, 133, 136, 137
Draine, Tom 46, 56
Drake, Cdr. Waldo 31, 32
Durgin, R.Adm. Calvin T. 156

E

Eggert, Lt. Joseph R. 86
Eldorado destroyer 156
Encyclopaedia Britannica 10, 16, 22
Engebi 72
Eniwetok 72, 74, 75, 81, 91, 93, 96
Enterprise carrier 144, 164, 171
Espiritu Santo 75, 76, 174
Essex carrier 37, 89, 90, 97, 103, 104, 113, 114, 117, 131, 133, 160, 164

F

Ferris, Philip L. 96
Fighter Director in Radar Plot 34-5, 39, 53
Fighter Director Officer (FDO) 153, 156, 163, 164, 168, 169, 170
Fighter Director School 40, 41, 45, 175
fighter squadron training at Beavertail Point, Quonset, RI 56-7
Fleet Radar Center see Camp Catlin
Flying 14, 15
Forbes, William A. 3, 38, 177
Ford, John 16
Formosa (Taiwan) 97, 98, 125-28
Forrestal, James, SecNav 43, 178
Franklin carrier 111, 151, 152, 154
Free French Navy 23, 24
Free French officers 23, 24
"friendly" or "bogey" identification 34-5

186   Angel on the Yardarm

G

Gambier Bay carrier 108
Gates, Artemus L., Asst. SecNav 38
Gatling destroyer 107
Gay, Ens. George 16, 18
Gehres, Capt. Leslie H. 152
Georges Leygues cruiser 65
Ginder, R.Adm. Samuel 70, 71, 75, 80
Golden Dragon Fraternity 74
Griffin, Cdr. Jack 33, 39
Grumman Aircraft factory 16
Grumman F6 Fighter plane 54-55
Grumman Hellcats 82
Guadalcanal 38, 75
Guam 85, 89, 91, 92, 106
"gulls," radar deception 126-27
Guthrie, Cdr. William 113, 141, 142

H

Haggard destroyer 154, 167
Halekulani Hotel, Waikiki Beach 30, 31
Halsey, Adm. William F. 70, 88, 93, 94, 95, 98, 101, 102, 104, 105, 106, 107, 109, 111, 117, 118, 121, 124-27, 141, 171
Hancock, Joy 9, 10, 11, 20-21, 31
Hannegan, Cdr. E.A. 55, 113, 114, 118, 125, 141, 142
Harlan, Lt. Herbert 25
Harrill, R.Adm. W.K. 86, 87, 91
Harvard Yard 5, 7
Hauser, John 46, 56
Henderson Field airport 38, 45
Hickerson, Loren 141
Hijo carrier 91
Hills, Hollis 82, 95
Hollandia 74, 80, 81
Hornet carrier 16, 18, 75
Howard, Lt. 36
Hoy, Lt. Ben 13

I

Independence carrier 42, 48, 165
Indianapolis cruiser 150
Ingalls, Cdr. David 30
Inland Sea 159
Intrepid carrier 68, 111, 114, 151, 153, 165
Iowa battleship 66
Irwin destroyer 107
Ise and *Hyuga* battleships 126
Iwo Jima 86-87, 91, 144, 146

J

Jones, Fred 65

K

kaiten see kamikazes
Kamikaze Special Attack Force 108, 155
kamikazes 111, 112, 114, 127, 134, 148-49, 150, 151, 156, 157, 161, 164, 165, 167, 168, 171, 172, 173, 174
  began at Leyte Gulf Battle 100; baka bomb 153, 154, 164; Betty bomber 151, 153, 154, 165; kaiten (human torpedo) 165
Karant, Max 14
Kavieng 74, 76
Kenmore transport 28
Kenney, Gen. George C. 80, 81
Kerama Retto 155-56, 158
Kiefer, Capt Dixie 117, 137
Kikusui (massive air attacks) I-X 157-61, 164, 167-68, 169, 171, 172
King, Adm. Ernest J. 9, 12, 19, 20, 21, 30, 171
Kinkaid, V.Adm. Thomas C. 101-09, 125
Knox, Frank, SecNav 9
Konrad, Cdr. Ed 71
Kurita, V.Adm. Takeo 101-03, 105-09
Kwajalein Island 70-72, 74
Kyushu 150, 151, 153, 155-58, 165, 168

L

Langley carrier 3-4, 43-52, 53-57, 59-65, 66-73, 75-79, 85-93, 102-05, 113, 114, 118, 121, 122, 124-127, pic. 135, 136, 138, 141, 146, 149, 156, 164-66, 169, 172-75
Lee, V.Adm. Willis A. 106
Levy, Max 55
Lewis B. Hancock destroyer 67
Lexington carrier 75, 80, 97, 103-05, 111
Leyte 94, 125
Leyte Gulf, Battle for 100-10
  Operation King II 100; Japanese Sho plan, three pronged 100-01; submarines in Palawan Passage 101-02; airstrikes from carrier groups 102-05; misleading reports of Kurita in Sibuyan Sea 105-06; Seventh Fleet to Surigao Strait 106-08; Halsey's decision 106-07; Kurita at San Bernadino Strait unopposed 107-09; carriers air response 109-10; Japan lost 305,710 tons of warships; U.S. lost 36,600 tons; hundreds of planes both sides 110
Lingayen Gulf 110, 114, 125, 127
Lingga Roads 127
Loomis, Lt. Lee 30
Lowe, Henry 98
Luzon 94, 125, 128
Lyman, Lauren D. (Deac) 3, 38, 177

M

MacArthur, Gen. Douglas 74, 80, 97, 100, 110, 111, 117, 118
Magnesium flares, enemy 77, 146-47
Majuro Atoll 71, 73, 78, 81, 84, 85, 96
Manus 74, 81
Marianas Turkey Shoot 89-91
Marine Corps 80n
May, Dick 81, 82
Meek, Samuel W. 38
Menjou, Adolph 20

Index 187

Meserve, Bob 46
Midway, Battle of 16, 88
Miller, Lt. Cdr. H.B. "Min" 13-14, 26
Mindanao 75, 94
Mindoro 117, 125
Mississinewa tanker 112
Missouri carrier 164
Mitchell, Grant 25, 26
Mitscher, V.Adm. Marc A. 70, 86, 88, 89, 90, 91, 100, 105, 111, 141, 150, 159, 171
Moffett, Adm. William A. 12, 22
Monosson, Adolphe 5-6
Monsarrat, Bonnie 37
Monsarrat, Marcus 37
Monsour, Edward 5
Monterey carrier 119
Morrison destroyer 104, 107
Mt. Suribachi 147
Mullins 175
Musa freighter 43
Musashi battleship 106, 159

Mc

McCain, R.Adm. John S. 20, 70, 102, 109, 110, 111, 141
McCampbell, Cdr. David 102
McCann-Erickson 75

N

Naval Air Transport Service 30
Naval Aviation 33-34
Naval Aviation News 26
Negros 94
Ngesebus 93
nightfighters used first time 81
Nimitz, Adm. Chester W. 28, 38, 70, 88, 94, 101, 108, 109, 159
Nishimura, V.Adm. 101, 107-08
North Carolina battleship 130

O

O'Callahan, Lt. 31, 32
Ofstie, Cdr Ralph 31, 86
Okinawa 91, 97, 98, 110, 126, 138, 147, 148, 150, 154, 155, 157, 158, 161, 167, 171-72
Olendorf, R. Adm. Jesse B. 106, 107-08
Operation Flintlock 72, 74
Operation Forager 85
Operation Iceberg 148, 150, 152-53, 172
Operation King II 100
Operation Reckless 80
Operation Stalemate II 93
Ormoc Bay 112
Osborne, Robert 13
Outlaw, Lt. Cdr. Eddie 78-79, 82, 84, 86, 90
Ozawa, V.Adm. Jisaburo 87-91, 101, 109, 159

P

Pacific Fleet Radar Center 32, 39
Pagan 86

Paine, Hal 87
Palau Islands 75-78, 80, 81, 93, 94, 95, 127
Panay 94
Parry Island 72
Patterson, Pat 87
Pearl Harbor 7, 25, 29
Peleliu 78, 93
Pepper, Lt. G. Willing 30
Philippine Islands 93-99
Philippine Sea, Battle of 85-92
　capture of the Marianas main objective 85; forces readied 85-86; the huge fighter sweep 86-87; opposing fleet 87-88; opposing enemy planes, Marianas Turkey Shoot 89-91; Spruance's plan 87-89; American submarines 90; night landings difficult 91; resistance ended on Tinian and Guam 92
Platt, Rutherford 3, 38
Platt-Forbes, Inc. 3
Pride, Capt. Alfred M. 19
Princeton carrier 75, 80, 97, 103, 104-05, 107, 110, 111
Pringle destroyer 165
public relations within the Navy 12, 13-14, 15
Pulitzer, Ens. Joseph 13, 31, 36, 37

Q

Quinhon 127

R

Rabaul 74
radar means radio detection and ranging 51
Radar Operator's Training School 52, 55-56
radar pickets in destroyers 112-13, 156, 158, 164, 169, 172
radar plot see Fighter Director in Radar Plot
radar repaired after typhoon 124
radar, shrouded in secrecy 33
Radford, Adm. Arthur W. 19, 130, 136, 144, 150, 163, 164
Randolph carrier 144, 149-50, 165, 171
Reed, Lt. Alan 104
Reeves, Lt. Donald E. (Dagwood) 82, 86
Reeves, R. Adm. J.M. (Blackjack) 22, 80
refueling dangerous 118
Reno cruiser 105, 111
Ridgway, Lt. C.D. (Chuck) 69, 86
Rota 85, 90, 91
Rounds, Ens. Frank 13, 31, 36
Rounsaville, Gus 45, 114
Royal Air Force (RAF) 3, 21, 22

S

St. Lo carrier 108
Saipan 85, 86, 89, 90, 92
Sallada, Capt. Harold B. 19, 23
San Bernadino Strait 105-08
San Jacinto carrier 117, 119, 136, 160
Saratoga carrier 75, 144, 147
Seiiche, V. Adm. Ito 159

Seiler, Lt. Eddie 96, 102-03, 138
Shaw, Artie 31
Shephard, Ens. Harry 46
Sherman, Capt. Forrest P. 30
Sherman, R. Adm. Frederick C. 93, 102, 107, 111, 112, 113, 117, 130, 136, 138, 144
Shikoku 150
Shima, V. Adm. 101, 108
Shokahu carrier 90
Sibuyan Sea 106
Smith, Linus (Linie) 42, 45, 47-49, 50, 52-57
Smith, Lt. Cdr. Tom 75, 113, 141, 174
Sorber, Tom 45, 47-52, 60, 67, 77, 124, 126, 149, 168
Soucek, Apollo 75
Southerland, Cdr. J.J. 170
Spence, Hull and *Monaghan* destroyers 121
Spruance, Adm. Raymond A. 70, 87-90, 91, 93, 106, 141, 150, 159, 171
Stefan, Nick 42, 46
Steichen, Edward 13
submarines, midget 112, 165
surface-search radar SG 33, 39, 40, 54, 60, 126, 154, 162, 167
Surigao Strait 106
Swanson, Lt. Cdr. Chandler W. 161

T

Tacloban 111
Taiho carrier 90
Taito 130
Takao harbor 130
Tama cruiser 109
Tang submarine 82-84
Taroa Atoll 71, 74
Tawi Tawi, Borneo 87-88
Thompson, Lt. Cdr. McKee 75, 113
Thorne, Lt. Cdr. Landon K. (Lannie) 33, 37, 38, 42
Thorne, Lt. Oakleigh 30
Threadfin and *Hackleback* submarines 159
Ticonderoga carrier 134, 136-38
Tinian 85, 89, 92
Tokyo 144, 146, 150
Tokyo Express 45
Tokyo Rose 98
Towers, V. Adm. John H. 3, 9, 11, 12, 13, 15-16, 29, 38, 70, 88
Truk 76, 81, 82
Tupper, Fred 13
Turner, Lt. Filo 55

Turner, Adm. Kelly 156, 158
typhoon 118-22

U

Uhlmann destroyer 154, 166
Ulithi 95-97, 111-13, 121, 138, 139, 140, 146, 147, 148, 149, 161-62, 168, 171, 174
United Aircraft Corporation 3
U.S. Navy, induction into 5-7

V

HMS *Victorious* carrier 37
Visayas 94

W

Wagner, Joe 96
Waldron, Lt. Cdr. John C. 16
Washington battleship 130, 131, 133, 138
Wasp carrier 45, 151-52
Wead, Cdr. Frank W. "Spig" 19
Wegforth, Capt. John F. 31, 95-96, 113, 124, 125, 133, 142-44
Wheeler, Tom 143
White, Cdr. Don 140-41, 146
Wickendall, Lt. M.M. 86
Wiltse, Capt. A.J. 29, 32
Wingard, Cdr. W.C. 141, 144
Winston, Lt. Cdr. F.L. 86
Woleai 78
Woodson, Lt. Walter B., Jr. 154, 166
Wordell, Cdr. M.T. 95-96, 109-10
Wotje Atoll 71, 74

Y

Yahagi 161
Yamato battleship 151, 159-61
Yap 78
YE Sector (note) 162-63
Yorktown carrier 69, 75, 80, 144, 150, 151, 165, 168, 169, 170
Young & Rubicam 75

Z

Zuiho cruiser 110
Zuikaku carrier 110